A Modern Introduction to Food Microbiology

BASIC MICROBIOLOGY

EDITOR: J. F. WILKINSON

VOLUME 8

A Modern Introduction to Food Microbiology

R. G. BOARD
University of Bath
School of Biological Sciences
Claverton Down, Bath

BLACKWELL SCIENTIFIC PUBLICATIONS

OXFORD LONDON

EDINBURGH BOSTON MELBOURNE

© 1983 by
Blackwell Scientific Publications
Editorial offices:
Osney Mead, Oxford, OX2 oEL
8 John Street, London, WC1N 2ES
9 Forrest Road, Edinburgh, EH1 2QH
52 Beacon Street, Boston
  Massachusetts 02108, USA
99 Barry Street, Carlton
  Victoria 3053, Australia

First published 1983

Printed in Great Britain by
Butler & Tanner Ltd, Frome and London

DISTRIBUTORS

USA
  Blackwell Mosby Book Distributors
  11830 Westline Industrial Drive
  St Louis, Missouri 63141

Canada
  Blackwell Mosby Book Distributors
  120 Melford Drive, Scarborough
  Ontario, M1B 2X4

Australia
  Blackwell Scientific Book Distributors
  31 Advantage Road, Highett
  Victoria 3190

British Library
Cataloguing in Publication Data

Board, R.G.
  A modern introduction to food
  microbiology.—(Basic microbiology; v.8)
  1. Food—Microbiology
  2. Food contamination
  I. Title   II. Series
  576'.163   QR115
ISBN 0-632-00165-8

Contents

Preface

This book on food microbiology is intended to be an introductory text for students of microbiology, food science, food technology and related disciplines. It has been assumed that such students will have received or be receiving an adequate introduction to microbiology *per se*. As emphasis was given at the planning stage to the introductory role of this book, it differs from most, if not all, current textbooks on food microbiology in giving emphasis to general principles rather than cataloguing information about the major categories of man's food. To provide a theme, food microbiology has been considered to be a facet of ecology, a concept that has pervaded both the theory and practice of the discipline in the 25 years since the classic review by Mossel and Ingram.

The current text is probably unique in its genre in citing, when appropriate, the day-to-day problems that have to be tackled by the practising food microbiologist. At the same time, an attempt has been made to set such examples against a wide perspective so that the would-be food microbiologist recognizes from the outset that success in his/her chosen profession depends on attention to detail as well as a thorough knowledge of the discipline and an awareness of the insults to which ingredients, as well as finished products, may be exposed. Moreover, the opportunity has been taken, when pertinent, to stress that the successful application of food microbiology in a factory is based on the collaborative efforts of persons from diverse backgrounds and training—the successful food microbiologist is simply a member of a team of successful specialists.

I should like to express my sincere thanks to Charlie Davidson, Martyn Brown, John Abbiss, Jeff Banks, Howard Tranter, Nick Sparks, Hilary Dalton, George Nychas and Chris Leads who through their research efforts have provided me with a deeper appreciation and understanding of the interplay of pure and applied microbiology, and to Mrs Rita Pratt who has managed quite spectacularly to type the manuscript from notes lacking calligraphic grace.

R. G. Board

Introduction

This book deals with only one of the many topics included under the heading, industrial or applied microbiology (Table 1)–food microbiology*. This table highlights an apparently immutable trait, the cataloguing of materials destroyed by microorganisms or products synthesized by them. This approach tends to compartmentalize the knowledge of the student and to isolate the specialists in one field of applied microbiology from those in others. Thus at a fairly early stage the study of applied topics can become divorced from that of fundamental aspects of genetics, biochemistry and microbiology. Although this introductory book surveys only one area, by emphasizing general principles rather than describing a few subjects in detail it provides a much broader introduction than the title might imply.

Table 1. Topics included in industrial microbiology.

Biodeterioration of	Timber and paper
	Paints and coating materials
	Metals
	Petroleum products
	Pharmaceutical products
	Wool
	Rubber
	Leather
	Paper
	Food and beverages
Biosynthesis of	Pharmaceutical products
	Foods and feeds
	Alcoholic beverages
	Industrial solvents and other chemicals
	Amino acids and other organic acids
	Industrial polymers, e.g. dextran
Reclamation of	Metals from low grade ores
	Water from domestic and industrial wastes
	Petroleum products other than by distillation
Assay of	Antibiotics
	Amino acids and vitamins in foods
	Industrial preservatives and disinfectants
	Pollution of the environment with fungicides, herbicides, etc.

What is the theme that can unify industrial microbiology? In the author's opinion it is the study of those factors which govern the growth of microorganisms. If in this approach, the substrate being attacked or the product being synthesized is considered as but one component in an environment, attention can be given to rate of attack or synthesis and thereby to the overall environment in which the changes are occurring. If, for example, one is dealing with the conservation of a food then the emphasis is given to the means whereby the rate of change due to microbial activity is minimized. On the other hand, a rapid rate of change is sought when microorganisms play an important role in the processing of a food, as with the fermentation of milk. This unifying concept of applied microbiology is discussed in Chapter 1 and provides the theme for the other chapters dealing with foods.

In selecting topics for discussion, care has been taken to select examples of ancient as well as modern methods. Such a choice emphasizes another feature of applied microbiology. Through the thousands of years that man has developed methods in which microorganisms are of cardinal importance, many tricks have often been included in a process and even today detailed analysis is required to identify those having a direct bearing on the process. Discussions of modern fermentation technologies highlight two features. A technology may well be based on the analysis of an old craft, as with the fermentation of milk and vegetables where the processes are based on those few facets which have a direct bearing on the performance of the fermentative organisms. In contrast, the processes used to produce single-cell proteins are based on a thorough knowledge of large-scale fermentation technology and the organisms tend to be selected for their capacity to exploit the optimal conditions offered by modern techniques.

These differences in approach are of relevance when one is considering the basic training required for a career in industrial microbiology. It is obvious that there needs to be an adequate coverage of biochemistry, genetics and microbial physiology so that the unifying theme noted previously has a firm theoretical basis. Allied to this is the requirement for skill in the enrichment, isolation and characterization of microorganisms and a sound understanding of the principles of ecology. These considerations dictated the choice of subjects discussed in the following chapters.

* For a detailed discussion of training for food microbiology see: Changing roles of food microbiologists in the 1980's, Feature articles (1982) *Food Tech.* **36**, 58–77.

1 Ecology and Food Microbiology

As noted in the Introduction, the absence of a common theme impeded the development of food microbiology as a discipline in its own right. The classic review by Mossel and Ingram (1955) provided a conceptual framework for the development of food microbiology in the past two decades. In practice the study of food preservation is the study of the means used to destroy or to inhibit the growth of those microorganisms which have the potential to cause deterioration of food. Three factors can be considered to be involved (Fig. 1.1):

1   *intrinsic factors*—the physicochemical properties of the food;

2   *extrinsic factors*—the physical and chemical properties of the environment, either bulk or micro, surrounding the food;

3   *implicit factors*—the physiological properties that enable particular organisms to flourish because of the interaction of (1) and (2).

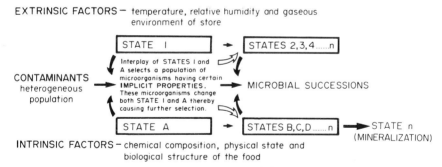

Fig. 1.1. Food spoilage and, eventually, its mineralization results from the selection of microorganisms having *implicit* properties which allow them to acquire numerical dominance (form a 'bloom') in the niches that develop from an interplay of the physicochemical properties of the food (*intrinsic factors*) and storage conditions (*extrinsic factors*). See Chapter 6 for further details.

There are two major but opposing strategies in the preservation of food. (a)   *The inhibition/destruction of microorganisms in or their removal from a food.* It has to be accepted that sterilization or aseptic production is applicable to particular foods only. The food microbiologist, therefore, becomes involved with the problems of hindering the growth and activities of organisms associated with the degradation and, ultimately, the mineralization of foods. It is important also to ensure that potentially pathogenic microorganisms do not survive in large numbers, multiply or leave toxic products in a food. A complete inhibition of microbial activity need not ensure stability of a food; chemical attack, e.g. fat oxidation, and degradation by enzymes, either of food or microbial origin, can lead to undesirable changes in flavour and texture (Fig. 2.5), processes which can be slowed by blanching (see p. 69), chilling, freezing, or the addition of antioxidants, for example, the addition of sulphite to apple slices in order to prevent browning due to polyphenol oxidases (Fig. 2.8).

(b) *The fostering of those organisms whose activities enhance and/or endow flavour, texture and colour; they may also inhibit the growth of potentially harmful organisms.* Today the successful outcome of many processes is helped by the use of specially selected microorganisms ('starter' cultures) or enzymes of microbial origin (Table 1.1).

Table 1.1. Microbial enzymes used in food manufacture.

Enzyme	Source	Use
Amylase	*Aspergillus niger* *Bacillus subtilis*	Manufacture of syrups
Glucose oxidase	*Aspergillus niger*	Removal of glucose from egg white that is to be dried
Invertase	*Saccharomyces cerevisiae*	Prevention of granulation in thick syrups and fondants
Pectinase	*Aspergillus niger*	Clarification of wines and beers. Extraction of fruit juices
Protease	*Aspergillus oryzae*	Prevention of 'chill haze' in beers
	Janthinobacterium lividum	Meat tenderization*
	Mucor pusillus	Curdling of casein in cheese production

* Used only in trials to date.

Setting aside the organization of the efforts of production workers, many of whom will be semi-skilled, by management, the well-being of a food in a developed country is determined by the application of principles by a relatively few technologists who factor the 'food' at some or all of the stages from harvest through production and storage until offered to the consumer (see Fig. 9.4). Indeed it is becoming more and more evident that little of value is achieved in food microbiology unless a complete production and distribution chain is considered.

To apply either of the strategies of food preservation noted on page 1, the conditions required for microbial growth must be understood. In the following section, some features of microbial growth will therefore be considered.

ENVIRONMENTAL GRADIENTS

In plant ecology, the concepts of environmental gradients are used to account for the distribution of species or communities along a latitude or longitude or from sea level to a mountain top. The emphasis is given to space because of its importance relative to geological time and evolution. With some foods, such as meat having easily defined portions of fat and lean, different communities or associations of organisms are selected because of the intrinsic, chemical properties of the two substrates. A layer of dilute syrup on the top of thick syrup is an example of a practical situation which would select facultative osmophiles (at the top) and obligate ones in the depths of the syrup. Thus, with fast-growing moulds and bacteria, there are situations in food microbiology when it is useful to consider the environmental gradient concept in the sense of 'a defined space for unlimited time', when attempting to predict or interpret the distribution of different microorganisms in a solid food or viscous liquid.

In many other situations, emphasis can be given equally to *time* and *space*. In other words, attention is directed at the *rate* at which a change occurs at a point within a food. This concept can be applied to: a food gaining or losing heat; a product in which a change in redox controls microbial selection and growth; the diffusion of salt, syrup or preservatives through a solid food—as with the diffusion of brine into meat; the production of acid by fermentative bacteria in products such as sauerkraut. Thus, for practical purposes, the organism introduced, either intentionally or unintentionally, to a food during the preliminary stages of processing can be considered to be 'stuck' at a point and, providing that the natural, implicit factors of the food will support its growth, it will be capable of growth for only as long as the imposed environmental stress is within its tolerance; for example, if temperature was being used as the environmental stress, the sluggish loss of heat could, say, result in a mesophile being provided with a temperature decreasing from 30 to 15 °C over five hours, during which time it might well achieve upwards of ten generations, whereas if the drift from 30 to 15 °C was achieved within three minutes, then no growth would occur (see Fig. 9.7).

Heat and milk

When attempting to achieve rapid heating or chilling, the bulk and form of the food to be treated is important as well as changes caused by processing. Thus with milk, the original method of pasteurization in a large tank (at 62.8 °C for 30 min) was replaced by the application of a thin film of milk to an efficient heat exchanger (71–73 °C for 15–17 s). Latterly, milk has been rendered germ free by treatment at 135–150 °C for up to 4 s, a process referred to as ultra-high temperature treatment (UHT). In practice UHT can be considered to be nothing other than a special case of a general approach to the heat treatment of foods (High Temperature Short Time (HTST)), where the objective is to achieve a germ-free state by the application of high temperatures (130 °C and above) for short times, i.e. a few seconds to 6 minutes.

Pasteurization of milk was adopted initially in the United Kingdom to kill *Mycobacterium tuberculosis*, but was subsequently used for short-term preservation. The UHT method gives 'commercial sterility' without inducing a cooked flavour in the milk because the temperature coefficient (Q_{10}) for change in reaction rate (Fig. 4.5) is greater for the destruction of the bacterial spore (Q_{10}, 8–30) than for chemical change (Q_{10}, 2–3). Although such heat treatments will kill the psychrotrophic flora that may develop in refrigerated milk, they will not inactivate completely thermostable enzymes secreted by the organisms. On storage, the processed milk can acquire a rancid flavour, due to lipases hydrolysing triglycerides, or bitterness, due to breakdown of casein by proteases, particularly if the numbers of micro-organisms in the raw milk were large. Indeed should such pasteurized milk be used for cheese production, the carry-over of thermostable lipases of microbial origin can cause rancidity during the ripening process. Thus although technological innovation allows the production of milk with a good shelf-like and no 'cooked flavour', an undesirable feature of milk in glass bottles sterilized in tanks of boiling water, it is successful only when the production and storage of the raw material have been rigorously controlled. Moreover, the success of a process such as UHT is only assured by aseptic dispensing into germ-free containers. As the latter are commonly formed from heat labile materials, novel methods of sterilization have been introduced so that the container is rendered germ-free without leaving a residue that could cause chemical taint. H_2O_2, UV light, Cl and hydrogen chloride have been selected for these reasons. Indeed, recent studies have indicated that when H_2O_2 and UV light are used in combination, the rate at which microorganisms are killed is greater than when either is used alone.

Tolerance profiles

Figure 1.2 illustrates the response of an organism to temperature, pH, salt or sugar concentration. Since it has been composed from data obtained from separate experiments using, for example, a psychrophile, a mesophile and a thermophile, the figure is probably spurious in a natural situation. From a theoretical viewpoint, it is probable that with foods containing a hetero-geneous flora, competition or synergism between organisms could give profiles of the type shown in Fig. 1.3. The growth of an organism can be controlled by the qualitative or quantitative deficiency or excess of one or more factors which approach the limits of tolerance of the organism. An

4

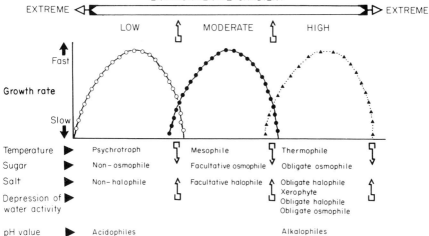

Fig. I.2. The profiles obtained in experiments in which pure cultures of microorganisms are used to determine growth rate in a medium containing a range of concentrations of H^+, salt, sugar, etc., or incubated over the termperature range from $c. -5-55\,°C$.

organism may have a wide range of tolerance of one factor, but a narrow range for another. When conditions for a species with respect to one extrinsic or intrinsic factor are approaching the limits of tolerance, the limits of tolerance may be reduced with respect to another ecological factor. In nature, and probably in foods also, an organism rarely lives at the optimum range as determined by laboratory studies of a particular factor, thus indicating that another, possibly unsuspected, component is growth limiting.

In the discussions of methods of preservation (see p. 21), examples are given of traditional foods (butter, cheese, bacon, etc.) whose stability has been shown to be dependent upon an interplay of factors, none of which was known to those who contributed to the evolution of the craft. Although current research is demonstrating the subtle nature of the interplay of pH,

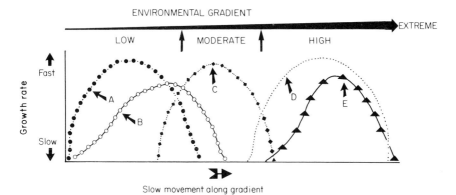

Fig. I.3. Hypothetical growth rates of contaminants in a food subjected to temperatures over the range $c. -5-55\,°C$ or changing concentrations of an additive such as sugar or salt.

5

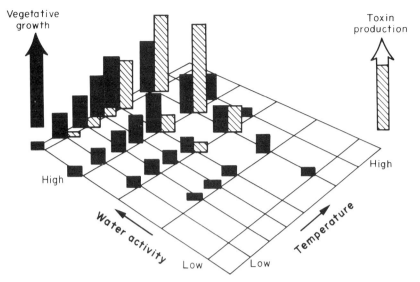

Fig. 1.4. The interplay of two factors on microbial growth in a food or a model system simulating a food are commonly displayed as 'sky-scraper' figures. This figure shows the influence of temperature and water activity on growth and toxin production by *Penicillium cyclopium* (redrawn from *Nestle Research News* (1980/81) Safety aspects of food fermentation, by M. von Schothorst).

NO_2^-, NaCl, pO_2, etc., in the preservation of ham and bacon, the complexities of the interactions are such that mathematical modelling (see Chapter 9) is only just beginning to replace the presentation of results as models based on three-dimensional histograms (Fig. 1.4).

Interplay of factors

It is from an appreciation of the factors involved in tolerance that food microbiologists are beginning to make foods whose stability under defined conditions of storage for known times can be assured. Thus, for example, a slight to moderate deficiency (water, pO_2) together with a slight to moderate excess (pCO_2, H-ions) and storage at -1–$4\,^{\circ}$C can provide a stability—and often a blandness—that would be difficult to achieve with one preservative alone. Moreover, the success of this approach to food preservation is dependent upon the investigators appreciating from the outset the probable nature of a spoilage flora. Thus, in a hypothetical case, through having its pH reduced to 5.5 and its water activity to 0.89 a product will support mould and yeast growth only and the former can be prevented by packing in an O_2-depleted atmosphere, either N_2 or CO_2. As leaks in packs flushed with gas can cause faults to develop during storage, gas packing was not readily accepted by industry until the introduction of 'cushion packs'. Gases at pressures greater than that of the atmosphere cause inflation of the packing materials; punctured packs are therefore recognized easily. The stability of grain having more than its alarm water content (see p. 31) is assured if the pO_2 in the void spaces is reduced either by sparging with CO_2 or by being denied exchange with the bulk atmosphere, the O_2 present originally being depleted by the activity of the grain and microbial contaminants. In the case of hay, drying in the field to low levels of water can be avoided and stability assured by adding fatty acids, e.g. butyric acid.

Although nomograms can be devised so that food manufacturers can choose appropriate combinations of low pH and sugar concentration for preservation, care must be taken in the uncritical application of such systems. Thus, the storage of olives in mineral acid instead of the fermentation products of lactic acid bacteria can cause spoilage by pectinolytic yeast, although the H ion concentration is about 1000-fold higher in the former case. In this and many other cases, the organic acid anion has itself an antimicrobial action, probably due to its lipophilic properties, which is more effective than the H ions themselves.

Detailed microbial analysis of a solid food which has had its intrinsic properties changed slightly and even one which has spoiled, results in the isolation of a wide range of organisms (Table 1.2). This suggests that such foods offer a variety of environmental niches or that the spatial relationship between contaminants, and thus the microcolonies formed by them, ameliorates the competition between organisms having similar implicit properties; alternatively, some type of synergism or symbiotic relationship between the organisms may be established such that a microbial succession is favoured or a stable mixed culture or consortium formed. It has been claimed that proteolytic pseudomonads may be essential pioneers in the succession leading to the dominance of non-proteolytic pseudomonads on meat and poultry. A complex example of synergism is provided by the fermented milk product, kefir, in which fermentation is brought about by lactic streptococci, leuconostocs and yeasts associated in large aggregates or 'grains'. If an environmental extreme, such as that resulting from relatively high quantities of salt or sugar, is used to preserve a food, the resident flora consists of a very few species, namely the obligative halophiles, osmophiles, or xerophytes. In other words, preservation can be considered to be a form of enrichment culture unless sterility is achieved (see Chapter 4).

Microbiological analysis

When attempting to interpret the information gathered from a detailed examination of a spoiled food which had not been exposed to some extreme of an environmental gradient, it is difficult to provide an explanation for those organisms which have not achieved dominance. Is their failure to grow merely a function of the inadequacy of the nutrients—either in variety or amount—in the food, due to antagonism by another organism or do they need to live synergistically with some other organism in the food? In recent years, the demonstration in the laboratory of one isolate from a food inhibiting

Table 1.2. The microorganisms associated with meat and meat products.

Process and product	Pseudomonas spp.	Acinetobacter spp.	Enterobacters	Clostridium spp.	Bacillus spp.	Micrococcus spp.	Staphylococcus spp.	Vibrio spp.	Brochothrix thermosphacta	Lactobacillus spp.	Streptococcus spp.	Leuconostoc spp.	Pediococcus spp.	Kurthia zopfii*	yeasts	Cladosporium herbarum	Moulds general
Joints																	
Freshly butchered	+	+	+	+	+	+	+	+	+	+	+	+	+	+	+	+	+
Stored at 0–4 °C in normal, moist atmosphere	++	+															
Stored at 0–4 °C in impermeable films—reduction of pO_2, increase in pCO_2	++	+						+	++								
Stored at 0–4 °C in atmosphere enriched with CO_2								+				++					
Stored at −1 °C in atmosphere enriched with CO_2										++							
Stored at 0–4 °C and surface allowed to dry						++								++		++	
Stored at 10–25 °C in normal, moist atmosphere	++	+			++					++	+			+	+	+	+

(continued)

Treatment							
Stored at >25 °C in normal, moist atmosphere	++	++	++				
Stored at >40 °C in normal, moist atmosphere		++	++				
Comminuted (ground) meat							
Stored at 0–4 °C in normal atmosphere	++	+			++	+	
Stored at –4 °C in atmosphere enriched with CO₂	+				+	++	
Containing sulphite					++		
Glucose added			++		++	++	
With biscuit crumb and sulphite			++		++	++	++
Cured meats							
Whole side of pork cured (NaCl, NO₃⁻, NO₂⁻) and heavily smoked—surface growth only			++	+++	+++	++	++
Pork cured (as above), lightly smoked, sliced and put into vacuum packs			++	+++	+++	++	
Pieces of pork cured, cooked, sliced and packed					++		

Table 1.2. (*continued*)

Process and product	*Pseudomonas* spp.	*Acinetobacter* spp.	*Enterobacter* spp.	*Clostridium* spp.	*Bacillus* spp.	*Micrococcus* spp.	*Staphylococcus* spp.	*Vibrio* spp.	*Brochothrix thermosphacta*	*Lactobacillus* spp.	*Streptococcus* spp.	*Leuconostoc* spp.	*Pediococcus* spp.	*Kurthia zopfii* *	yeasts	*Cladosporium herbarum*	Moulds general
Cured meats																	
Large pieces of meat cooked and then allowed to cool slowly					+ +	+ +											
Meat treated experimentally with tetracyclines															+ +		
Meat, with bones removed, placed in bag of impermeable film, atmosphere reduced in pressure ('vacuum pack') and stored at 0–4 °C	+ +									+ +	+ +						

* Although common in abattoirs and on meat, it rarely achieves numerical dominance on any product.
In all except freshly butchered meat, + + = extensive growth; + = moderate growth.

another from the same food has been often reported—but do the laboratory observations really apply in the food in question? It would seem that this evidence can be accepted in the case of liquids offering little impediment to the flow or diffusion of metabolites. Caution should be exercised with solid foods unless it can be demonstrated that the proximity of one organism to another is such that inhibition or enhancement of growth is feasible.

When considering the diversity of species in a food, it is essential to realize that many of the organisms may be simply remnants of associations present on an ingredient of the food or representatives of associations of habitats with which the food has had contact. In the latter case, they may be taken as indicators of the contamination of foods with members of an association in an environment with which the consumer ought not be linked by the food processing chain. Thus, by analogy to water, it has been customary to draw an unfavourable conclusion from the presence of *Escherichia coli* in food. Its presence indicates either direct or indirect contamination with the microbial associations found in the gut of man or animals, and hence their faeces. From a public health viewpoint, there has been the traditional and perhaps justifiable inference that *E. coli* may be associated with the presence of food-poisoning organisms (Table 1.3), particularly *Salmonella*. In the past few years, the studies of the ecology of plasmids has directed attention to their involvement in the transfer of drug resistance, for example, between bacteria, and has given another cause for the food microbiologist being concerned about faecal contamination. There is the fear that through serving as a vehicle in the spread of R-factors, a food may take antibiotic resistance and perhaps attributes of pathogenicity from one population to another (see Fig. 9.14).

Table 1.3. The food-poisoning bacteria.

Organism	Vehicles of infection
Salmonella spp.	Meat, particularly chicken meat; milk and cream, eggs
Staphylococcus aureus	Meat pies and meat products containing salt; confectionery
Clostridium perfringens	Cooked and reheated meats and meat products
Vibrio parahaemolyticus	Sea foods
Bacillus cereus	Rice and other starchy foods
Clostridium botulinum	Home-canned vegetables, etc.
Campylobacter jejuni	Chicken and milk
Yersinia enterocolitica	Pork and milk
Escherichia coli	Meat products

Table 1.4. Factors contributing to *Pseudomonas* contamination of curing brines (after Gardner G. A. (1980) *J. Appl. Bact.* **48**, 69–74).

Pseudomonas spp.	Contributing factors
Grow on pork during storage prior to curing	Length of storage and the temperature and relative humidity of the store will determine the rate and extent of their growth
Grow during the defrosting of frozen pork	Length of the defrosting period and the temperature will influence the development of the population of these organisms
Survive in heavily contaminated brines	A 90% reduction in the viable counts in brines (salt content, *c.* 26%) at 4 °C can take from 25–46 days

A recent study by Gardner (Table 1.4) provides an interesting example of a situation in which the monitoring of a stage in the manufacture of bacon provides clues about the microbiological status of the meat being processed. The presence of the salt-intolerant pseudomonads in bacon-curing brine ($>20\%$ w/v NaCl) is evidence of the use of meat on which the organisms have grown because of poor hygiene or poor temperature control, especially during the thawing of frozen pork.

The influence of the environment in which the plant or animal is raised or held before processing must also be considered. Thus psychrotrophic microorganisms make only a small contribution to the initial contamination of meat from animals raised in warm climates or fish caught in warm water. It has been noted that different types of acetic acid bacteria occur at different times of the year (Table 1.5) and, more particularly, at different stages in cider production. Thus species which preferentially oxidize sugars are found during the early stages of apple processing, the more acid-tolerant, alcohol-oxidizers at the time when yeast fermentation is almost complete. It has been noted recently that nearly all *Gluconobacter* and *Acetobacter* cause a type of rot and browning in apples and pears. Thus the changing pattern of isolations of acetic acid bacteria in fermenting apple juice (Table 1.5) may reflect in part changes in the incidence and level of post-harvest rots of apples due to infection with these organisms.

With gamebirds (order Galliformes), changes in the flavour of undamaged muscle of uneviscerated carcasses is due to autolysis, providing the conditions of storage and the diet of the bird do not favour growth of microorganisms in the gut. Thus it was noted that a gamey flavour developed during storage (9 days at 10 °C) of shot pheasants that had fed naturally on acorns and grain or had been fed grain supplemented occasionally with cabbage. There was no appreciable change in the size or composition of the microflora of the gut. Extensive greening of germ-free muscle was associated with abundant growth of *Escherichia coli* and *Clostridium* spp. in the duodenum and other parts of the gut of shot pheasants. These organisms produced

Table 1.5. Isolation pattern of acid-tolerant, Gram-negative bacteria.

Source	Month	No. of strains	*Acetomonas* spp.	*A. mesoxydans*	*A. aceti*	*A. zylinum*	*A. rancens*	*A. ascendens*	Others	
Cider manufacture										
Orchard soils	1	4							4	
Apple tree blossom	4	0								
Wild flowers	6	2	2							
Mummified apples	7	1	1							
Immature apples	8	0								
Fallen apples	8	8	3						5	
Factory equipment	8	0								
Apple tree leaves, twigs	9	4							4	
Mature apples	9	1	1							Pressing season (Mature apples to Fermenting juice, month 12)
Orchard grass, soil	9	0								
Factory equipment	9	8	4	2		1			1	
Harvested apples	10	63	47		8	3			5	
Harvested apples	11	3				3				
Pressed pomace (1st)	10, 11, 2	21	11	3		7				
Expressed juice (1st)	10, 11, 2	28	13	5	2	5			3	
Pressed pomace (2nd)	10, 11, 2	17	4	4		8			1	
Expressed juice (2nd)	10, 11	22	2	9	1	10				
Fermenting juice	12	2	1				1			
Farm cider	1	6	4				2			
Spoiled cider	1	12	5	2		1	4			
Dry cider	4	7	1	1		1	4			
Fermenting juice	5	1					1			
Factory sewer	6	14	2	3	1	1	5	2		
Other										
Cider vinegar generator	10	36		3	1	14	18			
Cider vinegar generator	11	10		5	1		4			
Mummified grapes	5	2	2							
Vines	6	1	1							
Strawberries	6	4	1				3			
Mature grapes	10	1	1							
Total		278	106	37	14	54	42	2	23	

copious amounts of H_2S *in vitro*. A similar discoloration of muscle has been noted also in pheasants fed proprietary turkey rations and killed by neck dislocation. Indeed the appearance of the muscle resembled that occurring in the spoilage of uneviscerated chicken carcasses where circumstantial evidence suggests that discoloration is caused by H_2S of microbial origin diffusing from the gut to the muscle. The carcasses of rock ptarmigan (*Lagopus mutus*) and willow ptarmigan (*L. lagopus*) are notable for their resistance to putrefaction when hung for long periods at temperatures up to 15 °C. This has been attributed to the former ingesting the preservative benzoic acid in cranberries (*Empetrum nigrum*) and the latter antimicrobial compounds present in the buds of willow and birch.

Not only has attention to be given, if possible, to identifying remnants of any associations present on ingredients or derived from contact with depots of infection (Fig. 9.4), it has to be considered whether or not such organisms have brought about, or their enzymes will bring about, chemical changes which may adversely influence some stage in the processing or storage of food. Thus the growth of acetic acid bacteria on apples before processing results in the production of 5-ketofructose and 2-ketogluconic acid, for example, which combine with, and thus neutralize, the preservative, SO_2. The growth of *Lactobacillus viridescens* in minced meat can result in the formation of peroxides (Fig. 2.6) which react with myoglobin (Fig. 1.5) to give green or colourless compounds and thus an obvious fault in products such as frankfurters or ham. Similarly, pectinases produced by moulds growing on decaying flowers of cucumbers can lead to maceration during the fermentation or storage of pickles. It was noted previously that lipases and proteases in pasteurized milk can cause changes in flavour and the latter can cause sterilized milk to gel during storage.

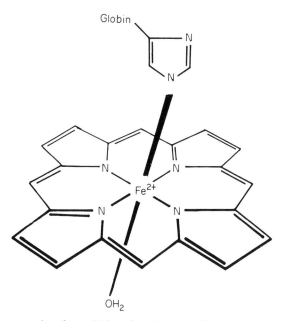

Fig. 1.5. The haem complex of myoglobin endows the meat of cows, sheep and pigs with its characteristic colour. The gaseous composition of a store or the addition of NO_2^- can change, either temporarily or permanently, the colour of meat (Fig. 2.9).

With the examples discussed above, opportunist saprophytes acquired during harvesting or during the initial stages of production of a food were the causative organisms of spoilage. In other situations, plant pathogens—the remnants of the flora on material infected in the field—can be the cause, the risk increasing when technology allows long-term storage of fruits or vegetables. In this context the plant hormone, ethylene (C_2H_4), is of importance because it can initiate the ripening of fruits picked in the green state. Thus the premature ripening of bananas—the skin changes from a green to yellow colour without a concomitant sweetening of the pulp—has been attributed to ethylene production in fruit infected with *Pseudomonas solanacearum*. If the rate of respiration of plant materials is retarded, then ripening or other physiological changes in the tissues are delayed and long-term storage is possible. A reduction of the pO_2 and an increase in the pCO_2 of storage (modified or controlled) atmospheres is one means of achieving this end (Figs 1.6 and 1.7). Alternatively a pressure lower than atmospheric (hypobaric storage) can be used because, in addition to reducing pO_2, it accentuates the rate of diffusion of C_2H_4 away from plant material.

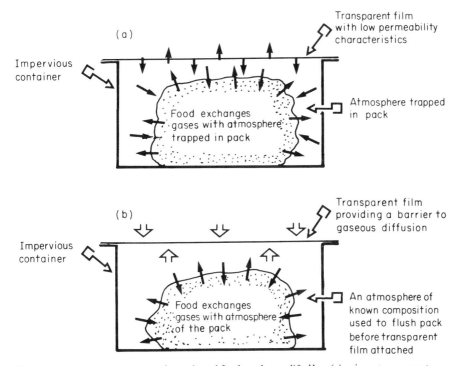

Fig. 1.6. The atmosphere around a packaged food can be *modified* by: (a) using a transparent film which impedes the diffusion of gas into or out from the pack, or (b) introducing a gas mixture into a pack covered with a gas-impermeable film. In either case an interaction between the packaged food and the trapped atmosphere can modify the latter.

Neither approach can be expected to influence adversely all plant pathogenic microorganisms; indeed the risk of plant material being damaged by such organisms is increased because of the extension in the storage life of the product. Recent research has indicated that the growth of pathogens can be controlled by including CO in a modified atmosphere. Thus the addition of CO to an atmosphere containing 2.3% O_2 and 5% CO_2 reduced by upwards of 90% the development of *Botrytis cinerea* on stored strawberries.

CARDINAL POINTS

When food microbiology is considered in the general context of ecology, it is evident that the concept of the environmental gradient provides a common theme for a discussion of many facets of the food industry. It has another useful feature also in that, when dealing with one of the environmental components of a food, some point along a gradient can be taken as an index in routine monitoring or control—examples are given in Table 1.6.

Table 1.6. The extremes of various environmental gradients that inhibit the growth of or kill selected organisms.

Temperature	
Minimum for growth of fungi	−8 °C
Tolerance of thermodurics	
e.g. *Streptococcus faecalis*	60 °C for 30 min
Microbacterium lacticum	72 °C for 15 min
Water activity	
Commonly occurring spoilage bacteria	
e.g. *Pseudomonas fluorescens*	0.97
Food-poisoning bacteria	
e.g. *Clostridium botulinum*	0.95
Staphyloccus aureus	0.86
Commonly occurring spoilage yeasts	
e.g. *Candida* spp.	0.88
Commonly occurring spoilage fungi	
e.g. *Alternaria citri*	0.80
Halophilic bacteria	
e.g. *Halobacterium*	0.75
Xerophilic fungi	
e.g. *Aspergillus echinulatus*	0.65
Osmophilic yeasts	
e.g. *Saccharomyces rouxii*	0.62
pH	
Clostridium botulinum	pH 4.6*
Zymomonas mobilis subsp. *pomaceae*	pH 3.6

* See Table 4.1.

Some of these have been derived empirically and others arrived at by experiment; for example, experience over the past twenty years or so has shown that few organisms grow on food stored at −7 °C and that none will grow at −10 °C. It will be appreciated that such statements tend to become 'a rule of thumb' in that they have a utilitarian purpose when applied generally—they rarely, if ever, link environmental gradients and ideas of tolerance.

In the preceding discussions, it was assumed that growth of organisms in food is essentially going to be under batch conditions, and for the majority of cases this is so. However, under certain conditions, a 'steady state' level of a limiting growth parameter can be reached when the rate of removal by microbial growth is equal to the rate of addition from some source, either by the action of another organism or by the involvement of a continuous process. Such conditions might appear to have limited application in food microbiology except in continuous fermentation systems such as the manufacture of single-cell protein and certain beers. It may be practical, however, with liquid foods such as soups, particularly if they need to be hot when put into containers, to establish a flow rate through a tank or tubes such that the flow rate is greater than the generation time of the organism so that it is unable to build up a population even though contained in a medium and at a temperature suitable for growth. Thus, through the design of equipment and planning of a process, the need for frequent stops for cleansing might be avoided by control of the flow rate. There would seem to be much scope for the application of this principle in food

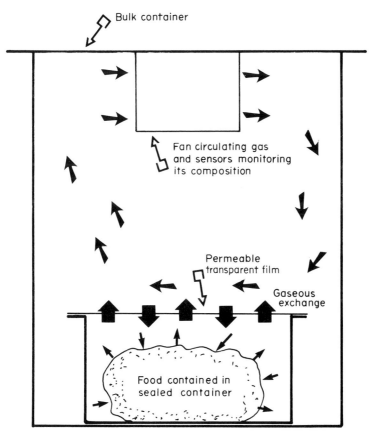

Fig. 1.7. The atmosphere in a packaged food can be *controlled* by using a transparent film which is freely permeable to gas and storing the pack in a closed container in which sensors assure that an atmosphere of a specific composition is maintained throughout storage.

microbiology providing, of course, that the surface of equipment did not trap organisms or the process did not select sessile organisms which colonized surfaces. In particular, it might find application in routine cleaning adopted for working surfaces if conditions could be imposed such that the generation times of the selected organisms could be extended to a period which would permit large-scale production in the intervals between cleaning.

Food processing

Although the concepts of ecology aid the organization and interpretation of data derived from studies in food microbiology, there is a major difference between the latter and ecology *per se*. With ecology, emphasis is given to the overall functioning of an ecosystem, its energy flow, the composition of its flora, etc. In food microbiology, utilitarian considerations demand that the major emphasis be placed on: (a) the control of contamination—the infection of a niche or habitat—so that spoilage is prevented or a desired fermentation is initiated, or (b) the hindering of the growth of spoilage or enhancement of growth of fermentative organisms. These objectives have to be achieved *routinely* in the industrial environment where the trend is always towards the large-scale production of individual units of food—the mass production of niches all of which may well have the potential to support large microbial populations. Although a food microbiologist will be concerned mainly with the day-to-day problems of manufacture, it is imperative that a food or that raw materials should be considered in relation to factors obtaining during breeding, growth, harvest, processing, distribution, storage, manufacture and use in the home or canteen (Fig. 9.4). Such an approach should lead to the adoption of methods which at all times complement the principal means of preservation. Indeed the present debate on the merits of Hazard-Associated Critical Control Points (HACCP) (see p. 51) is due largely to the recognition that little of value attends the microbiological analysis of a finished product when no attention is given to ingredients, methods of manufacture and storage conditions.

The preparation or processing of a food can and normally does change the physicochemical (intrinsic) properties of its main ingredients and perhaps the extrinsic properties of the bulk to such an extent that the spoilage association on the processed food differs markedly from that on the major raw materials. This is exemplified by the information given in Table 1.2. During the slaughter of animals—pigs, sheep, cows, chickens—a few mesophilic organisms of gut origin may be translocated to the tissues when the constitutive antimicrobial defence systems are negated. If the meat is not chilled rapidly, organisms such as *Clostridium* spp. may grow and cause putrefaction. In modern slaughter houses the eviscerated carcasses of domestic animals are cooled at such a rate that the growth of mesophiles and the tainting of the meat around the bones are prevented. The cooling rate needs critical control otherwise the advantages gained by the micro-biologist will be offset by the problems facing the food technologist (see

p. 46). As will be seen from Table 1.2, butchering leads to the surface of meat becoming contaminated with a wide range of organisms. With storage at 1–4 °C, the *Pseudomonas*/'*Achromobacter*' (*Acinetobacter*) complex, of which *Pseudomonas fragi* is the main organism, forms large populations on beef, poultry, pork and fish held in a normal atmosphere at a relative humidity which prevents drying of the surface. If, however, these organisms are prevented from growing, other contaminants will form large populations. Thus the addition of tetracyclines results in yeasts becoming dominant and drying out of the surface leads to the growth of moulds, e.g. *Cladosporium herbarum*. Wrapping beef, lamb, etc., in gas-impermeable film causes an accumulation within the pack of CO_2 of microbial and tissue origin. This limits the extent of growth of *Pseudomonas* and *Acinetobacter* spp. and *Brochothrix thermosphacta* makes a major contribution to the flora. This organism becomes the principal spoilage organism when, for example, beef is held in an atmosphere containing at the outset 20% (v/v) CO_2, providing storage is at 4 °C. With CO_2 storage at temperatures less than 0 °C lactic acid bacteria are the principal organisms in the populations developing on the meat. In this example, an extrinsic factor, temperature, is easily identified as the major elective agent. The information given in Table 1.2 indicates that at temperatures between 0–37 °C, various combinations of organisms (e.g. the pseudomonads and micrococci at 10 °C) dominate the spoilage flora. To account for these differences in the composition of the spoilage flora, it seems reasonable to assume that growth rate is the main implicit factor which determines whether or not an organism can contribute to a spoilage association. Such a view has received support from studies in which there was an obvious correlation between the growth rates of pseudomonads, *Br. thermosphacta* and lactic acid bacteria and the rates at which each of them spoiled meat.

Pseudomonads dominate the flora developing in minced beef held at 4 °C, thus indicating that grinding the meat does not lead to the redox falling to levels sufficiently low for the growth of clostridia. The addition of sulphites to minced meat elects a population dominated by *Br. thermosphacta* and this organism is the numerically dominant contaminant in British fresh sausages to which sulphite is added at levels of 450 $p/10^6$. Lactobacilli are the main contributors to the flora developing in minced meat to which glucose has been added and this situation is exploited by the manufacturers of 'continental' sausage, the stability of which is dependent in part upon acid accumulating during the fermentation process. Likewise, the stability of cured meats is dependent mainly upon NaCl electing the non-putrefying organisms: micrococci, lactobacilli, staphylococci, vibrios, enterococci and yeasts. Thus, the major factors inhibiting or encouraging the growth of particular microorganisms in foods can rightly and valuably be considered in the context of 'microbial ecology'. In the next chapter some of the major procedures used to influence that ecology and inhibit microbial growth are considered.

2 Inhibiting the Growth of Microorganisms

The principles of food preservation are easily listed:

1 prevention of damage of a food by mechanical causes, animals, insects, etc.;

2 retardation of chemical changes in a food through enzyme action or purely chemical reactions such as fat oxidation;

3 prevention or delay of microbial decomposition of a food;

4 the enhancement of growth of specific organisms which aid preservation through transformation of component(s) of the food.

PRACTICAL APPROACHES

There are three broad practical approaches to the problem of preventing or delaying microbial decomposition of food:

(a) inhibiting the growth of organisms;
(b) the removal of organisms;
(c) killing organisms.

The removal of organisms is a commercial possibility only with liquids such as wine, beer, cider, etc., where filtration is easily achieved.

If a traditional food or a range of ingredients is considered as a medium for microbial growth (Table 2.1), then in theory many strategies could be adopted to achieve preservation. In practice, of course, the choice is governed by economics, feasibility under commercial conditions and the need to cause minimum changes in the palatability or nutritional quality of a food. Moreover, with preservatives, the quantities used must be low so that they are non-toxic to man even when small amounts are consumed regularly over a lifetime. In practice many methods of preservation of cellular foods will result in damage to the cytoplasmic membranes; the chaos that can result from such damage is shown in Fig. 2.1. It has to be recognized that tradition, fads and prejudices impose restraints over and above those noted above. Indeed a food microbiologist has to attempt the application of fundamental knowledge in an area where at times quite arbitrary restrictions operate. Thus attempts to improve the microbiological stability of a traditional food will be unsuccessful if they modify, even slightly, one or other of the accepted organoleptic qualities of that food. The influence of tradition is seen in another context also; the terms drying,

Table 2.1. Preservation methods related to important components of a food or extrinsic factors.

	Method of preservation
Component	
Carbohydrate	Extraction and purification, e.g. sucrose. Fermentation to acidulate a food
Proteins	Extraction and purification, e.g. texturized vegetable protein
Amino acids	No commercial method available
Vitamins	Avidin's binding of biotin considered by some to be an important component of avian eggs' antimicrobial defence
Fats	Extraction, purification and emulsification, e.g. butter, lard and margarine
Trace metals	The binding of Fe^{3+} by ovotransferrin appears to be the major antimicrobial agent in the hen's egg
Water	Removal by drying or combination with sugars, salt and other humectants
pH	Deliberate acidulation. Fermentation of a carbohydrate
Extrinsic factor	
Gaseous environment	Removal of O_2; addition of CO_2
Temperature	Storage at $-1-4\,°C$ or at $-18\,°C$
Irradiation	Ultraviolet light

salting, smoking, pickling, for example, are used to classify a product that has received a particular treatment. In the following discussion it will be shown that in many instances the overall effectiveness of what on casual observation appears to be the main method of preservation is aided by the contributions of one or more minor factors. Indeed efforts to make certain foods such as pickles and salt meat more bland or to produce new systems have centred largely on modifications of the interactions that have been identified in traditional foods. The discussion will give examples also of foods in which a preservative contributes to the preservation of organoleptic properties by influencing chemical as well as microbiological changes.

Nutrient limitation

It will become evident when reading this chapter that rarely, if ever, is it possible to identify one component as the sole cause of preservation of a food; normally it is the outcome of a subtle interplay between a major and several minor components. Thus the choice of sub-headings in this chapter is intended to aid the cataloguing of information relating to the major components of preservation. It must be appreciated that anomalies will arise in such an approach. Thus in a discussion of food preservation through nutrient limitation it would be logical to consider water as a 'nutrient'

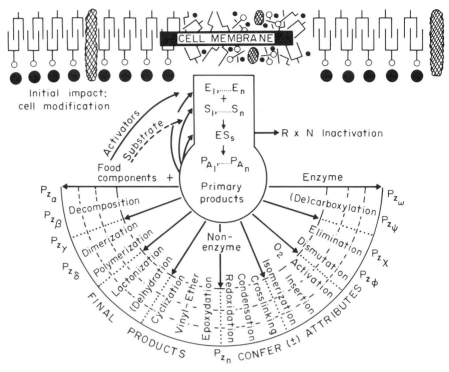

Fig. 2.1. Processing of cellular foods can cause major perturbations in enzyme activity resulting in damage to the cytoplasmic membranes (redrawn from Schwimmer S. (1978) Influence of water activity on enzyme reactivity and stability. *Food Tech.* (May 1980), 67).

and include dried foods or Intermediate Moisture Foods under this sub-heading. In practice such a course has not been followed because of the intention to direct the reader's attention to biological systems that have evolved to thwart microbial growth in fluids which contain abundant water and nutrients. The discussion of the antimicrobial defence of the hen's egg (see p. 24) deals with such a system. Indeed it is now becoming widely recognized that inhibition of microbial growth through iron deprivation, the major antimicrobial defence system in egg albumen, is a common strategy in biological fluids and medical scientists (e.g. Kochan 1973) have coined the term, nutritional immunity, for this form of defence.

BUTTER

With commodities such as sugar, edible fats of plant and animal origin, butter, etc. (Table 2.2), microbiological stability—but not necessarily chemical stability—is achieved mainly by reducing to a minimum extraneous substances, including water, which support microbial growth. With edible fats and sugars, extraction and purification is the task of chemical technologists. Butter, on the other hand, is a food whose production remains a part of the traditional food industry. Its microbiological stability is

Table 2.2. Foods preserved by nutrient limitation.

Fats of plant and animal origin
Butter
Margarine
Sugar*
Starches and gums

* Permissible water content of sucrose:

$$x = \frac{\text{weight of water}}{100 - \text{non-sucrose solids}} \times 100.$$

In very refined sugar with about 1% of materials other than sucrose, x should not exceed 0.03%.

dependent mainly on the residual non-fat components of the milk together with water used in processing being dispersed throughout the fat, the latter contributing about 80% and water about 17%, by weight, of the product. Churning and processing cause the water to be broken up into very small droplets. Some estimates suggest that there are $1-2 \times 10^9$ water droplets g^{-1} of butter. This implies that a large proportion of the droplets will have dimensions less than those of the bacterial cell and that raw materials would have to be excessively contaminated if the majority of droplets of more than 1 μm diameter were to be infected. It has been deduced, for example, that 88% and 99.9% of the droplets would be sterile in butter contaminated with 10^5 and 10^3 bacteria g^{-1}, respectively. If, as would be the normal situation, the contaminating flora was heterogeneous, then the number of droplets containing organisms capable of growth would be but a fraction of the total number of contaminated droplets. Even the largest droplets will contain only small amounts of nutrients and, as water constitutes the dispersed phase, the inward diffusion of nutrients and outward movement of wastes would be impeded. Although microbiological stability is dependent mainly upon processing removing most of the non-fat components of milk and dispersing the remnants throughout the butter, this cannot be relied upon when the product has to be stored for long periods. The control of contamination during manufacture and the use of pasteurized milk will obviously minimize the level of contamination of droplets with spoilage organisms. The environment afforded by the droplets can be made unfavourable by the inclusion of NaCl and/or lowering its pH by fermenting the cream before manufacture. Even these refinements will not ensure stability during extended storage and recourse has to be made to refrigerated storage, $0-4\,°$C for short and $-17--20\,°$C for long periods. It is common practice to store large blocks of butter and to prepare 250 g packs immediately before distribution at which time butters from various sources and of different grades may be mixed (blended). This can trigger off bacterial growth and hence spoilage. Why should this be when additional nutrients are not added? It would appear that blending causes both a redistribution and aggregation of droplets with the result that nutrients are supplied to organisms which had exhausted those available in the water surrounding them at the time of manufacture.

THE HEN'S EGG

The well-being of the embryo in a cleidoic egg such as that of hens is dependent upon a non-specific defence against microorganisms; the better understood defence based on antibodies and phagocytes would not work in the absence of a vascular system and neural control. The white makes an important contribution in terms of chemical defence (Table 2.3). In addition

Table 2.3. Chemical defences in the albumen.

Substance	Action
Lysozyme	Hydrolysis of β1-4 linkages in peptidoglycan of the cell walls of sensitive bacteria
Ovotransferrin	Chelation of Fe^{3+} leads to stasis or death of microorganisms
Avidin	Combines with biotin thereby making this unavailable to microorganisms
Riboflavin-binding protein	Combines with roboflavin thereby making this unavailable to microorganisms
Non-protein nitrogen	Only very small amounts available thus restricting microbial growth
Hydrogen and hydroxyl ions	pH 9.6–10.0; not only toxic *per se* but an adjunct in Fe^{3+}-deprivation by ovotransferrin

to lysozyme, for which no convincing case can be made for its contribution to the egg's defence, the white is an inadequate medium for microbial growth because of: (i) its low content of non-protein nitrogen, (ii) the unavailability of biotin and riboflavin through combination with proteins, and (iii) the avid chelation of Fe^{3+} by ovotransferrin. Of these, ovotransferrin working in association with the high pH (9.6) of the albumen, is of primary importance and its effectiveness is temperature dependent. At low temperatures (4–20 °C) microorganisms are prevented from growth by iron deprivation; at 38–40 °C, they are killed. Suggestions that organisms, such as coliforms, would be able to overcome iron deprivation by synthesizing their own chelates, enterobactin, have not been supported by laboratory studies. Indeed this was to be expected because enterobactin, even if formed in response to the iron-poor conditions of albumen, would be quickly hydrolysed and the resulting monomers have poor chelating potential. This defence plays a cardinal role in protecting table eggs from microbial attack providing that good storage conditions (controlled temperature and

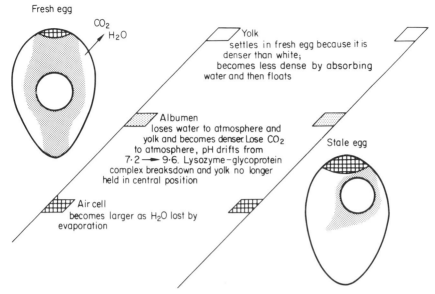

Fresh egg

CO_2
H_2O

Yolk
settles in fresh egg because it is denser than white; becomes less dense by absorbing water and then floats

Albumen
loses water to atmosphere and yolk and becomes denser. Lose CO_2 to atmosphere, pH drifts from $7\cdot2 \longrightarrow 9\cdot6$. Lysozyme–glycoprotein complex breaksdown and yolk no longer held in central position

Stale egg

Air cell
becomes larger as H_2O lost by evaporation

Fig. 2.2. Changes in the hen's egg during storage. The rates of change are directly related to storage temperature, the fastest changes occurring at high storage temperatures.

humidity) limit the rate of decay of the biological structure of eggs (Fig. 2.2). Indeed with storage at room temperature, the defence of the egg is dependent upon the yolk being kept away from the shell membrane, otherwise a niche is provided at the place of contact of the yolk and membranes for microbial growth in the absence of albumen. At chill temperatures ($0-4\,^\circ$C) rotting may well occur through chance collision of contaminants of the white with the yolk, the opportunity for collision being influenced by the size of the load of contaminants in the white. This load can be increased markedly if Fe^{3+} is deposited along with micro-organisms on to the shell membranes when eggs are washed. As this element remains localized, extensive microbial growth at the original site of infection results in heavy contamination of the albumen and rapid rotting.

COWS' MILK

Freshly drawn milk is an unfavourable medium for microbial growth. In addition to a few phagocytes, it contains transferrin/lactoferrin that chelate Fe^{3+}, agglutinins and the lactoperoxidase system which kills microorganisms through the oxidation of enzyme systems, with H_2O_2 as the substrate and thiocyanate an essential co-factor (Fig. 5.11).

pH

When it is recalled that certain heterotrophic bacteria can initiate growth in nutrient media poised at pH 1.0 (e.g. *Sarcina ventriculi*) or 9.6 (*Streptoccus faecalis*) and that there are microorganisms which have pH

25

Table 2.4. Foods in which pH contributes to microbial stability.

Acid added		Acid, mainly lactic, produced during microbial fermentations in
Acetic acid	Phosphoric acid/HCl	
Pickles—	Soft drinks	Sauerkraut
onions,		Olives
cabbage		Pickled cucumbers
Sauces		Cheese and other fermented milk products
Mayonnaise		
Tit-bits*		Sausages
Marinated herrings		

* A semi-preserved herring product.

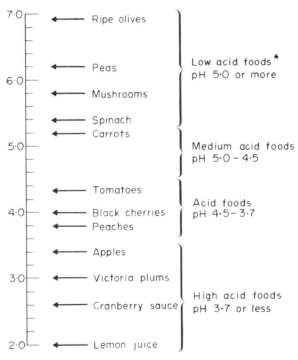

Fig. 2.3. The pH of vegetables and fruits used in canning.* The four levels of acidity that determine the choice of temperature/time relationships for appertization of such materials (see Chapter 4).

optima somewhere between these extremes, it is obvious that pH *per se* is unlikely to be a completely effective preservative. Nevertheless, pH is of importance as an adjunct to other methods of preservation (Table 2.4) or processing such as canning (Fig. 2.3). Moreover, pH is of particular importance from the viewpoint of food poisoning; until recently it was presumed that growth or toxin production by *Clostridium botulinum* was inhibited in a milieu having a pH of 4.6 or less. It has been shown that growth and toxin production can occur at pH 4.2, the critical pH value being determined in part by the acid used (Table 4.1).

FRUITS AND FRUIT JUICES

Many fruits have acid juices and, except for syconus ones such as figs, their contents can be regarded as germ free providing the integument is whole and healthy. When this barrier is destroyed, the juice may, and normally does, act as a selective agent in the enrichment of acid-tolerant organisms present on the fruit or processing equipment. Thus with apples used in cider production, few lactobacilli are present on the skin of hand-picked fruit, many can be recovered from fruit which have been collected from the ground on spiked rollers and enormous numbers are present in juice expressed by dirty processing equipment. The range of organisms associated with acid products are listed in Table 6.2. In certain cases, the lower level of pH tolerance of a spoilage organism may be higher than the commercially acceptable acidity of a product. Thus with *Zymomonas mobilis* subsp. *pomaceae*, the causative organism of cider sickness, growth is inhibited at pH 3.6–3.8. Cider manufacturers can, therefore, attempt to control this organism by blending apples so that bulked juice has a pH below these values. In addition, zymonads can be competitively inhibited by encouraging an active yeast fermentation and rendered quiescent in stored cider by ensuring that essential energy sources, glucose and fructose, are depleted by fermentation. The procedures adopted for controlling this organism must not, of course, favour the growth of other acid-tolerant spoilage organisms, such as the acetic acid bacteria.

In general, however, the acids occurring in a food or raw material will not ensure stability and other means of preservation have to be adopted, viz. appertization, pasteurization, gas storage, or the addition of preservatives (Table 2.10). The naturally occurring content of benzoic acid in the juice of cranberries is such that fruit submerged in water will keep for weeks.

ACIDS FROM MICROBIAL FERMENTATIONS

Although acid foods are not necessarily microbiologically stable, acidulation of a product having a neutral reaction may aid preservation (see examples in Table 2.4). Thus with certain plant materials (cabbage, olives, cucumber, etc.) which have juices rich in carbohydrate, poor in buffering capacity and of limited capacity to poise the redox potential, fermentation is used to increase the H ion concentration. In practice, the plant materials are held in tanks of brine, the salt acting both as a selective agent and an aid in the withdrawal of plant juices. With some material, such as olives, there may be and normally is a requirement to breakdown the antibacterial agents, aglycone and elenolic acid, which arise from the hydrolysis of oleuropein.

Oleuropein   β-3, 4-Dihydroxy-phenylethyl alcohol   Elenolic acid   Oleuropein aglycone

β-Glucosidase

During the fermentation of a product such as sauerkraut, there is a succession from populations of low (*Streptococcus faecalis* and *Leuconostoc mesenteroides*) through moderate (*Lactobacillus brevis*) to high-acid tolerant organisms (*Lactobacillus plantarum*). The succession of microorganisms in the depth of the tanks is interrupted at this point by a combination of acid, salt and anaerobic conditions. The combination of salt and low pH is perhaps the main reason for the failure of *Clostridium* spp. to contribute to the fermentation. Of course stability of the product will not be assured if acid-tolerant aerobic microorganisms grow on the surface of the brine at the expense of the acids formed by the fermenting flora. In practice, the growth of such organisms is controlled by UV-irradiation or by the exclusion of O_2 through sealing the top of the vats with water-filled plastic bags. When the fermented material is packed in small containers, the exclusion of O_2 is obviously important and stability can be ensured by pasteurization (see Chapter 5).

With the growing concern about environmental pollution and the consequent difficulties in the disposal of concentrated brine solutions, some manufacturers of pickled olives have attempted the bulk storage of material in water containing small amounts of organic acids and salt. Successful preservation can be achieved by storing olives in a solution of lactic acid (pH 3.8–4.10) containing 3–4% instead of the customary 6–7% (w/v) of NaCl. Stored olives spoiled when lactic acid was replaced by a mineral acid even though the pH was lower (pH 2.6) than that in the lactic acid solution; fermenting, pectinolytic yeasts caused softening of the olives and the integuments were distended in places by gas formation. Olives can be stored in a solution of acetic and lactic acid containing sodium benzoate providing the immersing fluid is covered so that aerobic organisms cannot breakdown the organic acids. These observations indicate that the acid radical rather than pH *per se* plays an important role in the storage of fermented plant material. Indeed this situation is now becoming widely recognized in food microbiology and not only in the context of food spoilage. Thus lactic and acetic acid, either singly or in combination, are more effective than hydrochloric or citric acid in inhibiting the growth of *Clostridium botulinum* and *Staphylococcus aureus*.

DELIBERATE ACIDIFICATION

The ·deliberate acidification of food represents a very old method of preservation, it having been applied to fish (marinated herring, soused mackerel), vegetables (onions and cabbages), pickles, sauces, and mayonnaise. Acetic acid has been the most commonly used acid. It is of interest to note that relatively few bacteria are tolerant of the acetate ion as witnessed by the successful application of acetate media in the isolation of *Lactobacillus*. The effectiveness of acetic acid as a preservative is dependent upon concentration as well as the pH of the food. Thus with pickles and sauces, acetic acid in a concentration of 3.5% of the volatiles ensures stability. Such a concentration gives a tartness which is becoming less acceptable to the consumer and, in a search for blandness and microbial stability, various combinations of acetic acid, humectants and preservatives have been tested. The admixture of sub-optimal concentrations of acetic acid and lactic acid have not proved successful in commerce. Adding sucrose and/or benzoic acid provides both stability and blandness. Moreover, the pasteurization of such products may at first sight appear to be merely additional insurance of stability. In practice, however, pasteurization may well be of particular importance, apart from destroying pectinases, in breaking a route of infection. Thus it can be assumed that the environment of factories in which sauces and pickles are produced will contain a flora characterized by the acetate tolerance of its members. Even if they did not spoil a sauce during storage and distribution, they might well do so at some time during use. As the general environment in the home or canteen is unlikely to be selective for acetate-tolerant organisms, the product would be at risk mainly to infection of factory origin. It is obvious that pasteurization is likely to play an important role in preventing such infection.

CHEESE

Cheese provides an example of a product whose stability is dependent in part upon the acid reaction resulting from the fermentation of lactose by streptococci and/or lactobacilli. It differs, however, from the fermented vegetables discussed earlier in that 'ripening' is associated with the biochemical activities of microorganisms, bacteria, moulds and yeasts, which succeed the 'starter' or pioneering organisms. Factors which enhance the succession are discussed subsequently (see Chapter 5). With cheese of the cheddar type, moulds can grow on the surface thus producing a fault as far as the consumer is concerned. This can be prevented by excluding oxygen, either by waxing the surface or covering it with gas-impermeable material, or by applying sorbic acid or one of its salts.

SAUSAGES

The microbiological stability of 'continental' sausages is in part dependent upon an acid reaction developing as a result of bacterial fermentation of carbohydrate added to comminuted meat. As with cheese, the 'ripening' of some such products is due to the growth of microorganisms which succeed the fermentative organisms, they may grow on the surface of the sausage, as with the moulds—mainly members of the genera *Aspergillus* and *Penicillium*—or within the sausage. In addition to the acids derived from the fermentation of carbohydrates, the preservation of fermented sausages is aided by an interplay of pH, the NO_2^- ion, salt, drying, smoking, etc.

RED MEAT AND pH

Red meats are considered to be less prone to microbial attack if they have an acid reaction. The latter results from the accumulation of lactic acid, the end-product of the glycolytic breakdown of glucose derived from the glycogen stored in the tissues. Glycolysis becomes the principal method of carbohydrate degradation immediately following the tissues' deprivation of O_2 due to exsanguination. The method of handling animals before slaughter has an important bearing on the pH attained in the tissues post-mortem. Excitement or fatigue will deplete the glycogen reserves; rest and feeding with carbohydrate will ensure that there is sufficient glycogen in the tissues for a reaction of c. pH 5.5 to be achieved post-mortem. In theory, an acid reaction in a cured meat such as ham or bacon would accentuate the toxicity of NO_2^- and, similarly, acid in a sulphited meat would enhance the bactericidal action of sulphite. In practice, however, modern methods of processing tend to produce products having neutral reactions due primarily to the use of 'polyphosphates' which improve the water-holding capacity of meat and meat products. Although claims have been made that phosphates, through acting as chelating agents, contribute to the preservation of foods, the available evidence does not permit an assessment of their contribution under commercial conditions.

Although the above discussion assumed that glycolysis results in a pH of 5.5 in all tissues, there are, in practice, differences in the pH obtaining in different muscles. This can lead to the localization of spoilage. With bacon stored under reduced pO_2, '*Pseudomonas mephitica*' causes greening/ blackening through H_2S production in those muscles with a reaction near to neutrality. Recent studies have shown that pH plays an important role in controlling the growth of *Brochothrix thermosphacta* on meat (Table 2.5).

Water activity

The amount of water in a food will obviously have an important bearing on its chemical and microbiological stability. Although this has been recognized for centuries and drying is one of the oldest methods of food preservation, little headway in understanding the principles of preservation

Table 2.5. The influence of pH and permeability of wrapping films on the growth of *Brochothrix thermosphacta* on meat at 5 °C (from Campbell R. J., Egan A.F., Grau F.H. & Shay B.J. (1979) *J. Appl. Bact.* 47, 505–9).

Conditions of storage	Populations (No. g^{-1}) formed at pH	
	5.4–5.7	6.0–6.4
Meat in closed tubes		
containing { oxygen	2.0×10^9	$> 1.0 \times 10^9$
containing { no oxygen	No growth	$> 1.0 \times 10^7$
Meat wrapped in transparent films of varying oxygen permeability		
low (1 ml)*	No growth	1.0×10^7
intermediate (150 ml)	c. 1.0×10^7	1.0×10^7
high (> 1000 ml)	1.0×10^9	1.0×10^9

* Oxygen permeability: value shown m^{-2} d^{-1} atm^{-1}.

was possible whilst attention was focused on the water content of a dried food. It was established that there is a fairly broad range in the alarm water content (i.e. the highest water content at which microbial spoilage will not occur) of traditional dried foods, viz. (% water in parenthesis): dehydrated whole egg (10–11%); wheat flour (13–15%); dehydrated fat-free meat (15%); dehydrated vegetables (14–20%); and dehydrated fruits (18–25%). Progress was rapid once the concept of water activity (a_w)— the ratio of the water vapour pressure over a food to that over pure water at the same temperature—was adopted. Although fundamentalists argue that it is the relative water vapour pressure (p_w) rather than a_w which is important, the latter has achieved common acceptance and will be used in the following discussion.

Some, perhaps the majority of bacteria, grow well in water containing small amounts of dissolved organic and inorganic substances. Others have a requirement for low concentrations of NaCl, for example the alteromonads; a facultative capacity to grow in high concentrations of salt, for example *Pediococcus halophilus* (growth in media containing c. 18% (w/v) NaCl), or an obligate requirement for solutions approaching saturation with NaCl, the halophilic bacteria. With yeasts, a large proportion of the known species will grow well in solutions containing 40% sugar; above this level osmophilic yeasts are selected and at sugar concentrations of 65–70% only a few yeasts can grow and then only slowly. There are some yeasts, for example *Debaromyces*, which grow in materials containing high concentrations of salt. It is generally accepted that moulds can grow under conditions which are too 'dry' for the majority of bacteria and yeasts, and xerophilic fungi are notable because of their capacity to grow under relatively dry conditions (Fig. 6.6).

When solutes are dissolved in water, some of the water molecules become arranged around solute molecules and there is an increase in the molecular forces between water molecules. This is reflected in a depression of the freezing point, a depression of water vapour pressure and the elevation of the boiling point. According to Raoult's law, the decrease in vapour pressure of the solvent of an ideal solution is relative to the mole fraction of the solute:

$$\frac{p_0 - p}{p_0} = \frac{n_1}{n_1 + n_2}$$

or that the vapour pressure of the solution relative to that of a pure solvent is equal to the mole fraction of solvent:

$$\frac{p}{p_0} = \frac{n_1}{n_1 + n_2}$$

when p and p_0 are the relative vapour pressures of the solution and solvent respectively, and n_1 and n_2 the number of moles of solute and solvent respectively.

From this derives the concept of the *equilibrium relative humidity* (ERH), the unique humidity at which the rate of evaporation and condensation are equal, viz. $\text{ERH}(\%) = a_w \times 100$. Moreover, in such a state the solution has a water activity (a_w) defined as:

$$a_w = \frac{p}{p_0}$$

Osmotic pressure is related to a_w, viz.

$$\text{osmotic pressure} = \frac{-Rt \log_e a_w}{\bar{V}}$$

where R = gas constant, t = absolute temperature, $\log_e a_w$ = natural log of a_w, and $\bar{V}$ = the partial molar volume of water. As reduction of a_w results from the increase in the concentration of a solution, it may be achieved by adding solutes—as in the preservation of foods by salting or syruping— or by removing water (Table 2.6). The refrigeration of a food will change

Table 2.6. Methods of drying foods.

Method	Foods to which applied
Insolation	Meat, fish and fruits
Smoking	Meat and fish
Accelerated freeze drying	Whole egg
Spray drying	Milk, egg albumen
Vacuum drying	Egg products

its a_w, the formation of ice crystals leading to a progressive increase in the solutes concentration. The relationship between water or moisture content (m) and a_w at equilibrium may be represented graphically by a moisture sorption (ms) isotherm (Fig. 2.4). Such an isotherm is normally sigmoid in shape. There are reasons for believing that, in practice, the sigmoid curve is formed from 3 'local isotherms' (li), shown as A, B and C in Fig. 2.4, which reflect three types of bound water:

1   monolayer—bound or oriented water;
2   multilayer—chemiabsorbed water;
3   mobile—capillary, solution.

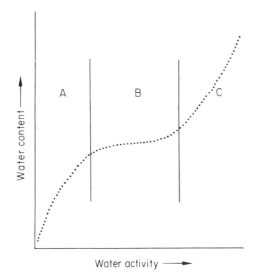

Fig. 2.4. The water sorption isotherm of a dried food.

From a microbiological viewpoint the water activity of a substrate having the local isotherm A, will not support the growth of microorganisms and, unless protected by being in the spore state or through being clothed with extraneous proteins, fats, etc., they may well die. In terms of chemical stability, the food will deteriorate due to autoxidation (Fig. 2.5) because the decay of free radicals (Fig. 2.6) is retarded by low moisture content. In the local isotherm, B, the a_w is such that it will be selective for xerophilic fungi (e.g. *Xeromyces bisporus*—minimum a_w for growth, c. 0.60/25 °C), osmophilic yeasts (e.g. *Saccharomyces rouxii*—minimum a_w, c. 0.62) and obligately halophilic bacteria (e.g. *Halobacterium*—minimum a_w, c. 0.75).

With the local isotherm, B, chemical changes (Fig. 2.5) will be due mainly to non-enzymic browning reactions (Fig. 2.7) or to enzymes (Fig. 2.8), activity of the latter increasing as increased moisture content aids the diffusion of substrates, providing diffusion is not impeded by biological structure. Thus with a 'natural storage product' such as grain having a

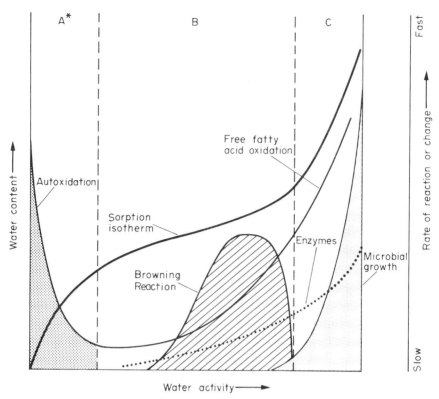

Fig. 2.5. Chemical changes and the rate of microbial growth in a food are determined by water activity.

* Local isotherms', for definitions see p. 33.

moisture content of *c.* 13%, there is negligible chemical deterioration even when it is stored for years. With flour produced from such grain, chemical deterioration can assume serious proportions in a matter of weeks even though the moisture content of the flour is equal to that of the grain prior to milling. In the local isotherm, C, the conditions will become less selective as the a_w moves from 0.80–0.98. In practice, the selective pressure will move from an a_w of 0.80 which will inhibit the growth of organisms such as staphylococci to an a_w of 0.98 which will allow the growth of pseudomonads, etc. Over and above its elective property, water activity will influence the rate at which microorganisms grow (Fig. 1.2). Thus a lowering of water activity away from the value which is optimal for a particular organism will lead to a progressive reduction in an organism's growth rate.

So far, we have been concerned with water activity from the viewpoint of a selective agent in the ecological setting. What is the relationship between water activity and the physiology of particular organisms? With bacteria such as the Archaebacteria, *Halobacterium*, which have an obligate requirement for high concentrations of salt, it has been demonstrated that they are uniquely adapted for life in such an extreme environment. Thus

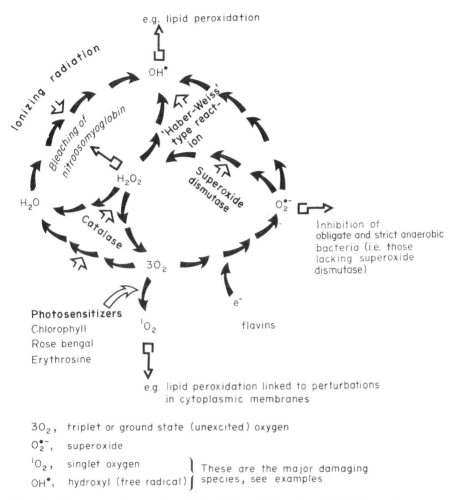

e.g. lipid peroxidation

OH•

'Haber-Weiss' type reaction

Ionizing radiation

Bleaching of nitrosomyoglobin

H_2O_2

Superoxide dismutase

H_2O

Catalase

$3O_2$

$O_2^{•-}$

Inhibition of obligate and strict anaerobic bacteria (i.e. those lacking superoxide dismutase)

e^-

Photosensitizers
Chlorophyll
Rose bengal
Erythrosine

$^1O_2$

flavins

e.g. lipid peroxidation linked to perturbations in cytoplasmic membranes

$3O_2$, triplet or ground state (unexcited) oxygen

$O_2^{•-}$, superoxide

$^1O_2$, singlet oxygen

OH•, hydroxyl (free radical)

These are the major damaging species, see examples

Fig. 2.6. The presence of oxygen in a food can lead to the formation of the highly reactive entities, singlet oxygen and the hydroxyl ion, that cause undesirable changes in texture and flavour. Figure composed by Dr A.D. Dodge.

their enzymes can only function when suspended in high concentrations of salts, their ribosomes are unstable when deprived of concentrated solutions of salt containing K^+, and their cell walls, which through lacking peptidoglycan differ markedly from those of eubacteria, require salt for stability. With salt-tolerant rather than salt-dependent bacteria, the evidence suggests that there are perhaps two major physiological groupings. There are those, such as some alteromonads, which have apparently an obligate requirement for a certain minimum level of Na^+, K^+, etc. There are others which have a marked tolerance of NaCl solutions giving a_w of c. 0.80–0.90 but which can grow equally well in a medium in which the a_w is poised by a mixture of salts other than NaCl. Some facultative halophiles or osmotolerant bacteria, e.g. *Bacillus* spp., achieve compatibility between

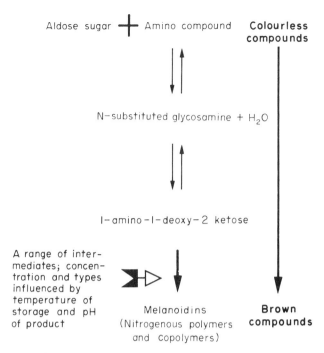

Aldose sugar ╋ Amino compound   **Colourless compounds**

N−substituted glycosamine + H_2O

I−amino−I−deoxy−2 ketose

A range of inter-mediates; concentration and types influenced by temperature of storage and pH of product

Melanoidins
(Nitrogenous polymers and copolymers)

Brown compounds

Fig. 2.7. Non-enzymic browning causes undesirable changes in the appearance, texture and flavour of dried foods.

the osmotic pressure of a medium and their cytoplasm by synthesizing compounds, proline and γ-amino butyric acid, that confer osmoregulation without disturbing the basic metabolism of the cell. Such compounds are secreted rapidly when a medium is diluted. With yeasts, the majority of which can grow with a minimum a_w level of 0.88, there are those which can tolerate sugar solutions of 40% and those which grow, albeit slowly, in sugar solutions of 65–70% (a_w c. 0.60). There are two main adaptations that fit certain yeasts, for example *Saccharomyces rouxii*, to growth in a medium of low water activity. In response to diminished a_w they synthesize polyols, mainly glycerol and a small amount of arabitol, which serve not only as osmoregulators but protect enzymes from damage due to low a_w. The cell membranes of the osmotolerant yeasts retain the polyols within the cell, whereas those of non-osmotolerant yeasts are leaky and much of the osmo-regulators produced in response to a diminished a_w is lost to the surrounding medium. Indeed the failure of some yeasts to become osmotolerant is probably linked to their inability to produce sufficient energy for cell maintenance and the synthesis of polyols in quantities sufficient to act as intracellular osmoregulators.

Although reduced a_w inhibits microbial growth it can confer thermal stability on microorganisms, a factor that needs to be considered if pasteurization of a dried product is contemplated.

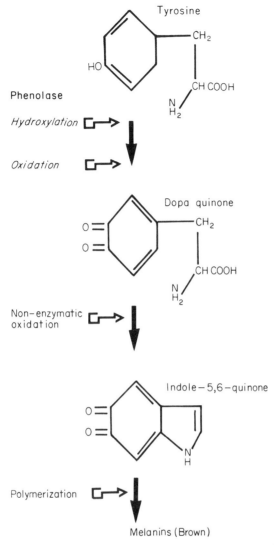

Fig. 2.8. Enzymic browning can mar the appearance of foods, viz. the browning of the cut face of apples and potatoes.

SALT

When considering the use of salt (NaCl) in a food, there is a need to define the role that it is playing. It may be merely a condiment; it may be bringing about some essential change in the physiochemical properties of a food—the extraction of plant juices in vegetables which are fermented with lactic acid bacteria, or the solubilization of the myofibrillar proteins of meat so that they may act as emulsifying agents during the preparation of a commodity such as Frankfurters—or it may be acting exclusively as a preservative, as in the case of dried salt fish (Table 2.7). With the last

Table 2.7. Foods in which NACl contributes to preservation.

Contribution as preservative	Food
As principal preservative	Salted fish
	Salted meat
As an adjunct to	
pH	Butter
	Cheese
	Fermented vegetables
NO_2^-	Bacon

mentioned there is a need to consider the actual concentration of NaCl in the liquid phase rather than in the food as a whole. Thus with an example discussed previously, salt will be concentrated in the water droplets in butter and it can be present at levels of 10% (w/v). The term 'brine' is used to denote the percentage of NaCl in the water phase of a food, viz.

$$\% \text{ Brine} = \frac{\% \text{ NaCl}}{\% \text{ NaCl} + \% \text{ water}} \times 100.$$

When considering the preservative action of NaCl, attention needs to be given also to the possible synergistic action of NaCl and other intrinsic factors, such as pH, or extrinsic factors, temperature, pO_2, etc. Thus the interplay of NaCl and pH is considered to be an important preservative system in cheese. Similarly, the salt in fermented vegetables is probably the principal agent in controlling the growth of clostridia.

CURED MEATS

From a historical standpoint, salting, or the curing of meat can be considered as a means whereby our forefathers preserved meat obtained from animals which had to be slaughtered in the autumn when fodder was no longer available. The Dutch used to call November, slaght-maand (slaughter month). In the evolution of salt curing, it was observed that contamination of salt with saltpetre led to the production of a heat-stable pigment, nitro-sylmyoglobin (Fig. 2.9), which imparted a desirable colour to the product. The development of methods of husbandry which provide a continuous supply of animals to the meat industry and the general adoption of refrigeration have reduced the need for a fully preserved product (Fig. 9.8).

At the outset, salt would presumably have been rubbed into the surface of meat and thus an inward diffusion of NaCl established. Although crystalline salt would provide an inimical environment at the surface, there would still be an opportunity for salt-intolerant bacteria of gut origin to grow within the tissues, particularly if there was inadequate chilling

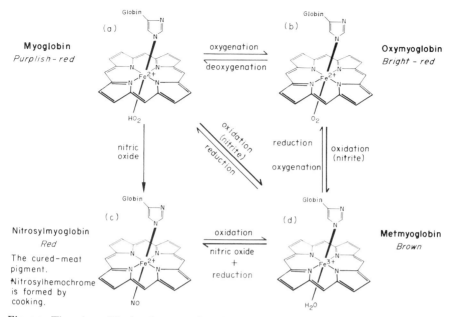

Fig. 2.9. The colour of fresh red meat products is determined by the compounds outlined above. See Dryden & Birdsall (1980) for illustrations.

post-mortem or storage was at a high temperature. In such a case, the growth of *Clostridium* and *Streptococcus* spp. around the bone in a ham— i.e. at the point farthest away from the salted surface—can produce the fault, 'bone taint'.

Once sufficient salt has diffused into the tissues, the hams or sides of bacon were smoked thereby drying the outer surface and distilling upon it chemicals, such as phenols, which retard fat oxidation, or formaldehyde, methanol, cresols, etc., which inhibit microbial growth. With this example, the 'case-hardened' surface will provide an environment inimical to most microorganisms other than moulds and a barrier against microbial invasion of the underlying tissues. A similar situation obtains with smoked fish and the efficacy of the external barrier is enhanced because proteins extracted by salt are denatured.

Instead of applying salt to the surface, the meat can be immersed in chilled brine (20–25% NaCl) containing NO_3^- and NO_2^- as in traditional methods of Wiltshire curing. The brines may be used for many years. Thus there is an opportunity for the enrichment of halophilic and halo-tolerant bacteria; halophilic Gram-negative rods which lyse rapidly in distilled water can form populations of 10^8 organisms ml^{-1} and the halo-tolerant cocci upwards of 10^6-10^7 organisms ml^{-1}. In addition, such brines contain lactobacilli, which are unusual in that they are intolerant of the acetate ion, and causal contaminants such as *Escherichia coli* and *Pseudomonas* spp. which are introduced with the meat (Table 1.4). The coliforms are of interest because they can retain viability for longer periods in chilled brine than they do in that held at ambient temperature. The microflora developing in brines are also of interest because of their negligible capacity to digest polymers of meat, thus they do not cause putrefaction. With halophiles such as *Vibrio costicolos*, NO_3^- is reduced to NO_2^-; not only is the latter toxic to many bacteria, it is involved in the reactions leading to nitrosylmyoglobin formation, hence the cured meat colour (Fig. 2.9). It also contributes to the flavour of bacon and ham. In fact, until the role of NO_2^- was recognized at the beginning of this century, curing was dependent upon the reduction of nitrate by meat and microbial enzymes and even today the 'cured' colour of certain fermented sausages results from microbial reduction of NO_3^-.

With the recognition of the important role of NO_2^-, the requirement for NO_3^--reducing bacteria could be avoided simply by adding NO_2^- to the brine or meat to be cured (Fig. 9.8). Likewise, the impregnation of tissues with salt and NO_2^- can be achieved by direct injection and a short period of storage in a 'cover' brine which may be used for a limited period only—a practice which may lead to a population dominated by halotolerant cocci rather than halophilic rods.

Cooking NO_2^--cured meat products may result in the formation of extremely small amounts (parts per billion) of nitrosamines (e.g. dimethyl-nitrosamine, diethylnitrosamine, dipropylnitrosamine, dibutylnitrosamine, etc.).

$$H_3C-N(H)-CH_3 + NO_2 \longrightarrow H_3C-N(N{=}O)-CH_3$$

Dimethylamine $\longrightarrow$ N-nitrosodimethylamine

Although the levels are low, the production of carcinogens during food manufacture or preparation cannot be ignored. As the derivation of nitrosamines is due to NO_2^- in meat, a reduction in concentration of the preservative offers one practical approach to the problem. Such a policy has to be balanced against the good public health record of cured meat products, especially those that rely upon a low heat treatment for stability. It has been established that heating of cured meat products accentuates the product's inimical properties in respect to *Clostridium botulinum*. Thus the reduction in the concentration of NO_2^- might conceivably enhance the risk of food poisoning due to this organism. A possible solution would be to diminish a meat's content of NO_2^- and add another preservative that acted synergistically with this anion. Promising results have come from trials with sorbates (see Chapter 9).

The slicing and packaging of bacon creates ecosystems different from that obtaining on smoked sides of bacon; as the former supports large populations of bacteria (micrococci, staphylococci, yeasts, lactobacilli and vibrios)—mould growth is prevented by vacuum packing in a gas-impermeable film—it can be regarded as a semi-preserved product only.

It is noteworthy that with this modern method of producing bacon, the salt still selects a flora that is unable to digest the polymers, especially the proteins, of the meat. Thus the product is acceptable to the consumer even though it may be harbouring relatively large numbers of bacteria and it is safe because of the failure of food-poisoning bacteria, especially enterotoxin-producing strains of *Staphylococcus aureus*, to compete with the other components of the flora particularly with storage at *c*. 4 °C.

SUGARS AND SYRUPS

The high sugar contents of materials such as maple syrup, honey, molasses, etc., mean they are stable unless contaminated with osmophilic yeasts or xerophilic fungi, providing, of course, that the available water is evenly dispersed. If storage conditions result in water sorption or, as may happen in a closed container, chilling causes condensation with water droplets falling on to the surface of the material, then conditions may well permit the growth of yeasts whose requirement for a a_w are greater than those of the true osmophiles. This can be a problem when liquid sugar is pumped into storage tanks. After the sugar has had its content of yeasts reduced by filtration through diatomaceous earth filters, it is cooled and pumped into steam-sterilized tanks, the head space of which is flushed with sterile warm air to prevent condensation and localized dilution of the syrup. UV lamps in the headspace may also be used to prevent yeast growth.

The stability of syrups of moderate strength can be improved if invertase is used to increase the concentration of glucose relative to sucrose (Table 2.8). Man has exploited this means of preservation in the storage of candied fruit, the fillings of chocolates and confectionery. Although control of microbial growth can be achieved by adjusting the concentrations of solutes in syrups, manufacturers often pasteurize such products so that they are freed of contamination with osmophilic yeasts. Thus with honey, preservation is assured by pasteurization (71–77 °C for a few minutes or 93 °C for seconds).

Table 2.8. Foods in which sugars aid preservation.

Syrups and candies
Fondant fillings in chocolate*
Honey
Jams and conserves
Candied peel
Dates, sultanas and currants

* *Fondant fillings*: $>79\%$ of sucrose and invert sugars prevent fermentation by osmophilic yeasts.

In the wine industry an interplay of sucrose and ethanol, which is 4–5 times more inhibitory than the sugar, in yeast fermentation of wine musts is measured in Delle units (Du):

$$Du = x + 4.5y$$

where x = g of reducing sugars 100 ml^{-1} and y = ml of ethanol 100 ml^{-1}.

Drying

The methods used for the desiccation of foods have been cited previously (Table 2.6) and the importance of moisture content on both the chemical and microbial stability of dried products discussed. In some instances, preservation can be achieved by linking a reduced moisture content with a modification of an extrinsic factor. Thus the successful preservation of cereals having moisture contents of more than the alarm water content of 13% can be achieved by storage in sealed containers where an increase in pCO_2 inhibits mould growth. Likewise, mould growth on cakes can be inhibited by enrichment of the storage atmosphere with CO_2 or the use of a fungistatic agent such as sorbic acid.

Intermediate moisture foods

There was little prospect of innovation when the stability of a dried food could be predicted only in terms of its alarm water content. The adoption of the concept of a_w led to: (i) the establishment of the minimum a_w required for the growth of a wide range of spoilage and food-poisoning microorganisms (Table 2.9); (ii) the realization that a_w was only a part of the preservative system in many traditional foods (dried fruits (a_w 0.72–0.80), jams (0.82–0.94), honey (0.75)), and (iii) the recognition that by use of appropriate humectants (salt, sucrose, sorbitol, propylene glycol) a

Table 2.9. Minimum a_w for the growth of selected organisms.

Bacteria	Min. a_w	Yeasts and fungi	Min. a_w
Eubacteria			
Pseudomonas spp.	0.97	Candida utilis	0.94
Acinetobacter spp.	0.96	Rhizopus stolonifer	0.93
Enterobacter aerogenes	0.95	Trichosporon pullulans	0.91
Bacillus subtilis	0.95	Aspergillus glaucus	0.70
Escherichia coli	0.96	Saccharomyces rouxii	0.62
Staphyloccus aureus	0.86	Xeromyces bisporus	0.60
Archaebacteria			
Halobacterium	0.75		

food could have its a_w modified without recourse to traditional methods of drying, syruping or salting.

Moreover, the demonstration that the minimum a_w for the growth of an organism is influenced by intrinsic (preservatives, pH) and extrinsic (gaseous environment, temperature) factors led to the possibility of 'modelling' a complex preservative system such that a long storage life at ambient or refrigerated temperatures could be contemplated.

The humectant can be incorporated in the product by: (i) soaking or cooking the food in a solution of the humectant; (ii) soaking a dehydrated food in a humectant; or (iii) regarding the humectant as an ingredient and mixing it with others in the preparation of a food.

The actual effectiveness of the required a_w will be influenced by the pH of a food and its content of preservatives. Thus with the addition of sorbate to control yeast and mould growth, an a_w of c. 0.8 would assure microbial stability. An acid food containing an antibacterial agent can be protected from microbial spoilage by an a_w of 0.85. Although the theory underlying the production of intermediate-moisture foods is relatively simple, practical problems have limited their uses, e.g. special foods required for manned space flights and pet foods. With the latter, a_w is controlled by a mixture of sugar and propylene glycol and shelf-life is aided by pasteurization. Unacceptable organoleptic properties are the major impediments to large-scale production of intermediate-moisture foods for humans. Besides texture and flavour problems due to the humectants available to food manufacturers, the a_w of such foods (0.70–0.90) is in the range (Fig. 2.5) in which non-enzymic browning, lipid peroxidation and enzyme activity can occur. Fat rancidity can be hindered by anti-oxidants and enzyme activity by heat treatment. As yet, however, there are few reliable remedies for non-enzymic browning.

The microbiological stability of a dried product often calls for critical control of the storage temperature. It is well known that moisture migration due to wide fluctuations in temperatures can produce 'wet spots' in which microorganisms grow.

Preservatives

In the Report (1972) on the review of Preservatives and Food Regulations 1962, the following definition of preservative was recommended: 'Any substance which is capable of inhibiting, retarding or arresting the growth of microorganisms, of any deterioration of food due to micro-organisms, or of masking the evidence of any such deterioration'.

It is obvious that a very wide range of substances could be used to achieve these ends. In practice, of course, the choice will be limited to those for which there is toxicological evidence that their use does not pose a threat to the health of man. There are other reasons for imposing additional restrictions on the use of preservatives. Thus a preservative should not be used if the microbial stability of a product could be assured by other means, viz. appertization, pasteurization, refrigeration, etc., providing that these do not impose prohibitive costs or give unacceptable organoleptic properties to a food. Moreover, a preservative should not be used merely to remedy faults arising from poor process control or bad hygiene at some or all of the stages in the storage, preparation and distribution of a food. Examples of preservatives are given in Table 2.10; the use of some of these is discussed in the sections concerned with a_w, pH, refrigeration, contamination control, etc.

Recent studies of British fresh sausages have shown that the preservative, sulphite—measured as SO_2 and permitted at an initial level of 450 p/10^6—plays many roles. It selects a microbial association in which Gram-positive bacteria (*Brochothrix thermosphacta*, yeasts and lactic acid bacteria) predominate and it inhibits the growth of enterobacteria. Comparisons of microbial growth and activity in sulphited and non-sulphited sausages have demonstrated that the preservative curtails by upwards of 1 log cycle the size of the climax populations of members of the association and influences their metabolism so that the pH of sulphited sausages drifts from c. 7.0–6.0, whereas a drift of 2 pH units or more is a feature of unsulphited sausages. In addition the preservative aids colour retention and protects to a limited extent the depolymerization of polysaccharides added to sausage; the combined action of amylase and maltase produces lower concentrations of glucose, maltose and maltotriose than they do in sausages lacking preservative. It is notable, moreover, that sulphite causes these effects even though its occurrence in the free state diminishes rapidly with time (Fig. 2.10). The major loss is associated with binding with materials in the sausage, the most likely binder being acetaldehyde of yeast origin.

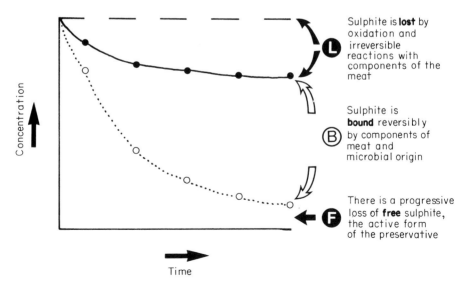

Fig. 2.10. The preservation of British fresh sausage is dependent upon free sulphite, the amount of which diminishes with time. The rate of binding of sulphite is temperature dependent, slow binding occurring at chill temperatures and fast binding at room temperature.

44

Table 2.10. Food preservatives.

Substance	Used in	To control
Sulphite	British fresh sausages	Growth of Enterobacteriaceae Acid production by microbial associations in sausages
Sulphur dioxide	Fermenting fruit juices	Growth of 'wild' yeasts
	Stored fruits	Growth of fungi, especially those that produce pectinases
	Dried and dehydrated vegetables	Colour loss
Nitrite	Meat products	Growth of spore-forming bacteria
Diethyl carbonate* (diethyl pyrocarbonate)	Fruit juices, wines and carbonated beverages	Growth of fungi
Ethylene or propylene oxide	Spices	Contamination
Acetate	Pickled vegetables	Microbial growth
Benzoate	Cheese and fruit juices	Growth of fungi and yeasts
Esters of *p*-hydroxybenzoic acids (Parabens) methyl ethyl propyl	Fruit juices	Growth of fungi and yeasts
Propionate Sorbate†	Confectionery, cheese	Growth of fungi
Formaldehyde	Cheese milk	Growth of *Clostridium* spp.
Nisin	Canned foods	Growth of spore-forming bacteria
Pimaricin	Cheese and sausage	Growth of fungi
Nystatin	Bananas	Growth of fungi
Chlortetracycline	Fish	Growth of bacteria

* Of limited use today; dimethyl carbonate may be widely used in future as a 'cold sterilizing' agent of fruit juices, wines, etc.
† See Overview (1982) *Food Tech.* **36** (December), 87–118 for recent studies of the control of growth of *Cl. botulinum*.

Temperature

For practical purposes, 3 storage temperatures for food preservation can be recognized: sub-ambient temperatures (cellar storage), chill storage, $-1-4\,°C$, and frozen storage, $-18\,°C$ or less. If the temperature variation along a 'cold chain' could be minimized, then 'super chill' storage at $-5\,°C$ would be a useful commercial system for storing perishable foods. The actual choice of temperature will be determined in part by the nature of the food and in part by the length of storage required. Thus with an example discussed previously, long-term storage of butter is achieved at $-17--20\,°C$, whereas chill temperatures are used for the storage and distribution of small packs. With fresh eggs having a small air cell, storage at temperatures below freezing can cause shell fracture due to ice formation in the albumen. Even if the shell is not damaged, disruption of the diffusion gradient at the surface of the yolk will cause Fe^{3+} diffusion into the albumen, the colour of which will become salmon-pink due to Fe^{3+} saturation of ovotransferrin. In other situations low temperatures may cause irreversible damage of a product through breakdown of cell membranes, as with bananas, or unacceptable changes due to enzyme action. Thus chill storage of potatoes causes glucose to be released from starch and, if they were intended for crisp production, the caramelization of glucose would cause discoloration. The blanching of vegetables is intended to destroy enzymes that cause changes in flavour during frozen storage (see p. 69).

Modern slaughter house practices, particularly refrigerated storage, have minimized the risks inherent in cooling meat too slowly. Indeed the practices are such that psychrotrophic microorganisms grow at the surface of the meat only (Table 1.2). With beef, for example, slime production and off-odours are associated with these organisms achieving populations of 10^7-10^8 organisms cm^{-2} of meat surface (Fig. 2.12). Not only will rapid chilling prevent 'bone taint' and delay the manifestation of spoilage at the surface of the meat, it will also minimize evaporation and hence the loss of weight of a carcass. Rapid chilling minimizes also the amount of liquid which drains ('drip loss') from jointed meat. Thus the food microbiologist and the accountant will always hope to achieve rapid chilling of meat, especially of beef and lamb. Too rapid chilling (the meat achieves a temperature of less than $10\,°C$ within 10 h) can result however in irreversible contraction of the muscles such that the meat is rendered tough ('cold shortening'). This is not an important problem with pork because the onset of *rigor mortis* is fast compared to that of beef and sheep carcasses. With the last two, 'cold shortening' can be avoided by controlling the rate of chilling (no part of the carcass should attain temperatures below $10\,°C$ within 10 h of slaughter) or the rate of development of *rigor mortis* through electrical stimulation of the carcass following removal of the hide and viscera.

Broadly speaking, cellar storage is used mainly for ware vegetables and fruits such as apples. Under such conditions, humidity may have to be controlled otherwise plant tissues will wilt or mould growth occur. Some ware vegetables, for example swedes, may be waxed before storage so that water loss is retarded. With some fruits, the natural defence imposed by the integument can be augmented with fungistatic agents. Thus with citrus fruits and bananas, for example, thiabendazole is applied to the skin after harvesting with the object of controlling mould growth. Similarly, the use of diphenyl in wrapping or packaging material is effective in reducing wastage of citrus fruit by mould growth. It has been shown also, that macrolide antibiotics such as nystatin, rimocidin and pimaricin are effective fungistats when sprayed on to soft fruits such as raspberries or strawberries. The fact that some of these antibiotics are medically useful in the treatment of 'thrush' tends to limit their use in the food industry. Current studies of vegetable storage are directing attention at the commercial feasibility of imposing controlled atmospheres (Fig. 1.7), either a reduced atmospheric pressure or an atmosphere of specific chemical composition.

CHILL STORAGE

When considering chill storage it is useful to recognize two situations: chill storage as
(a) the principal means of preservation;
(b) an adjunct to other means of preservation.

From the discussion of the selective influence of processing of meat (Table 1.2), it was noted that a reduction in storage temperature imposes a selective pressure on the heterogeneous population of contaminants acquired by meat during butchering. This results in psychrotrophic organisms being enriched by refrigerated storage. From a consideration of growth rate, it could be anticipated that temperature-induced retardation of growth will extend the shelf-life of a product. Such a view has been supported by studies in which a linear relationship between growth rate and the rate of spoilage has been demonstrated (Fig. 2.11); in other words, shelf-life has been linked directly with an implicit property of microorganisms, their rate of growth as influenced by an extrinsic property, temperature. With beef, it has been established that spoilage—slime production and off-odours—is associated with populations of 10^7-10^8 microorganisms cm^{-2} of meat surface. When consideration is given to the initial level of contamination and the shelf-life of a product, it has been noted that these two factors are inversely related (Fig. 2.12). The shelf-life will be determined ultimately by the proportion of organisms in the initial contamination capable of growth during refrigerated storage. Thus the success of such storage will be aided by high levels of hygiene obtaining at all stages of production and the absence of niches within the factory wherein psychrotrophic organisms can grow and spill over on to the product.

It is obvious that reliance on chill storage will impose serious limitations on the shelf-life of a product such as meat. In attempts to overcome this, many attempts have been made to modify the environment offered by

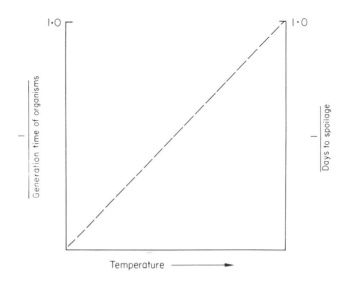

Fig. 2.11. The rate at which meat in a moist atmosphere spoils is determined by an interplay of the generation times of the spoilage organisms and the temperature of storage.

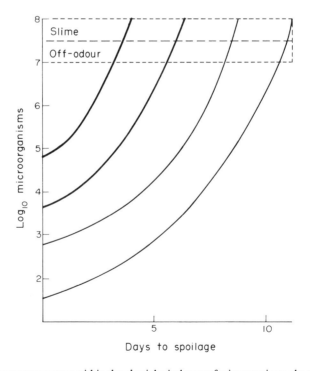

Fig. 2.12. At any temperature within the physiological zone of microrganisms, the rate at which meat spoils in a moist atmosphere is a function of the initial level of contamination with spoilage organisms.

packaged meat so that microbial growth is further retarded. In one method extended shelf-life was achieved by storing meat in a modified atmosphere containing 30% (v/v) CO_2 and 70% O_2. This inhibits the growth of members of the *Pseudomonas–Acinetobacter* group and the meat is colonized eventually by the slower growing *Brochothrix thermosphacta*. Likewise, an extension of the storage life of fish in ice can be achieved by adding antimicrobial agents to the ice.

DEEP FREEZING

Deep freezing (temperatures below $-10\,°C$) provides a means of long-term storage of food providing that the biological structure of the food is not destroyed and its texture or flavour made unacceptable by enzyme action. Contaminants on the food at the time of preparation are not killed—at least, the majority are not—so that there is probably a need to provide clear instructions so that the consumer may select conditions which allow the product to thaw without encouraging growth of microorganisms, particularly food-poisoning ones such as *Salmonella*.

The success of food storage at sub-ambient temperatures will depend ultimately on the quality of temperature control, not only during the time that the product is in warehouses but during transit to the shop, and home. This is referred to as the 'cold-chain' and, for success, every effort has to be made to ensure that the temperature, be it $-20\,°C$ or $1\,°C$, is maintained at all stages of distribution. Perhaps through ignorance or disregard for some of the principles of physics, chilled or frozen foods can be subjected to many abuses during storage and distribution. Indeed the siting of a thermometer in a refrigerated store can be of importance. If it is situated in the air stream coming from the condenser, then the recorded temperature may well be an index of the efficiency of the machine rather than the temperature obtaining within the food in the storage compartment. Likewise it is possible to entertain the wrong notions about the intended purpose of refrigerated containers. The size of the refrigeration unit and the quality of the lagging of the storage compartment may be adequate to ensure the maintenance of a particular temperature during the distribution of a food providing that the food was at or below this temperature at the time of packing. It will be appreciated that special problems are posed by galleys on commercial aircraft because limitations on the weight of equipment often preclude mechanical refrigeration. Alternative systems—refrigeration based on solid CO_2 (Drikold or dry ice) or freon—are designed so that chill temperatures are maintained rather than *attained* during flight. With such a system, it is imperative that the food is at the desired temperature when loaded on to the aircraft. This can be assured by storing the food for 3 or more hours in a *large* cold store before putting it in the galleys. Much larger refrigeration units and more efficient lagging would be needed if the intention was to reduce the temperature of the food during transit. Indeed, the effectiveness of the cold chain would

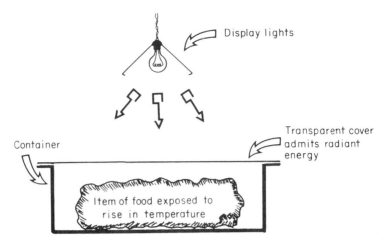

Display lights

Container

Transparent cover
admits radiant
energy

Item of food exposed to
rise in temperature

Fig. 2.13. The 'greenhouse' phenomenon. The storage of packs with transparent lids in a chill cabinet situated beneath bright display lights can cause the contents of the packs to heat up.

In future, a very thin coating of a reflective material, such as aluminium, on the lid may be used to reflect heat without impairing perceptibly the transmission of light.

probably be improved if refrigerated containers were rated for their potential to maintain (not achieve) specified temperatures for defined periods in the climatic conditions in which they were operated. Many examples of bad practices in the storage and display of chilled or refrigerated foods may be noted in shops. Thus the display of deep-frozen products in transparent wrapping materials can lead to the greenhouse phenomenon (Fig. 2.13), especially if bright lights are situated immediately above the display cabinet. Likewise a goblet-shaped container (Fig. 2.14) for chilled foods may be an inappropriate shape if the maintenance of a chill temperature depends upon a refrigerant circulating through hollow shelves.

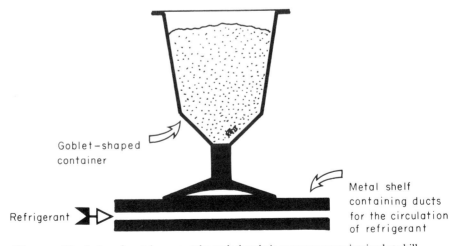

Goblet–shaped
container

Refrigerant

Metal shelf
containing ducts
for the circulation
of refrigerant

Fig. 2.14. The design of containers must be such that their contents are maintained at chill temperatures throughout storage. The stalk of a goblet-shaped container may effectively lag the contents when refrigerated shelves are used to display merchandise.

3 Control of Contamination

A major objective in the production of foods must be the control of contamination with food-poisoning organisms (Table 1.3) or their toxins (Table 9.3), and organisms that are physiologically equipped to spoil a food under normal conditions of storage and distribution. With the last mentioned, the possibility of enzymes remaining after the death of contaminants must be considered also. Thus in attempts to control contamination, attention must be given to depots of infection, the dispersal of organisms, and ingredients that may have supported large populations of microorganisms. Indeed, these three features must be covered by the policy that needs to be established so that all the personnel associated with a food factory contributes to the control of contamination. The phrase Good Manufacturing Practices (GMP) has been coined to cover an integrated approach to the control of quality and contamination of foods. In practice GMP will be based in large part on specifications or codes of practice, the effectiveness of which will need to be monitored regularly. Thus the microbiologist needs to select appropriate methods for the quantitative isolation of specific groups of organisms or their metabolic products. The day-to-day application of such methods must be influenced by statistical considerations relating to the probable distribution of contaminants (random *v.* contagious) and the inherent degree of risk of a food spoiling during distribution or, more importantly, acting as a vehicle in the transfer of pathogens, or their toxins, to susceptible hosts. In practice, therefore, the microbiologist will be establishing, or adopting microbiological standards, specifications, guidelines or bacteriological reference values for routine use. Such specifications must accomplish what they are purported to achieve and they must be technically and administratively feasible with relatively modest laboratory facilities.

METHODS

The choice of appropriate microbiological methods is essential if the results of a monitoring programme are to be of value in the day-to-day management of a factory. With the isolation of microorganisms, for example, the media and temperatures of incubation have to be selected with clear objectives in mind. Is the object to determine the extent of contamination of a food with food-poisoning organisms, the remnants of an environment with which the food ought not to have had contact, or the organisms which are physiologically equipped to cause spoilage? It would be fatuous to use Violet Red Bile Agar with incubation at 37 °C in attempts to monitor the level of contamination with spoilage organisms of a food which was to be stored at chill temperatures. It would, of course, give information on the level of contamination with coliform organisms and, therefore, providing the source of the ingredients and methods of production were known, some indication of conditions obtaining during manufacture or processing. With pasteurized milk, the recovery of coliforms would be evidence of post-processing contamination. To monitor the population of organisms capable of causing spoilage, a medium such as Plate Count Agar with incubation at 22 °C may well be appropriate providing, of course, some intrinsic factor of the food had not selected, or could be expected to select, organisms having particular nutrient requirements. Thus if lactic acid bacteria were sought, a rich medium (containing 0.5% (w/v) of peptone, meat extract and yeast extract) having an acid reaction (pH 5.4) and containing 0.2M acetate buffer would be used.

From a study of general microbiology, the essential features of a range of appropriate media will be known from those sections dealing with systematic bacteriology and mycology. There is one further feature that must be considered. It has been established that contaminants of foods can be

Table 3.1. Injury and repair of microorganisms. For additional details see Busta F.F. (1978) *Adv. Appl. Microbiol.* **23**, 195–201.

(a) The spores and vegetative cells of microorganisms in foods can be injured by exposure to

Elevated temperatures	Preservatives
Temperatures just above freezing	Acid or alkaline conditions
Freezing and thawing	Low a_w due to drying, freezing
Irradiation	or humectants

(b) Injury is manifested by failure of microorganisms to

Form colonies on nutrient agar that supports the growth of uninjured cells
Form colonies on nutrient agar containing a selective agent, e.g. deoxycholate, that is tolerated by uninjured cells
Break out of the lag-phase of growth

(c) Injured microorganisms can be resuscitated by

Incubation, at least initially, at sub-optimal temperatures
Incubation in a non-selective medium before transfer to a selective medium or in a non-selective medium into which a selective agent is released slowly from a capsule
Addition of a specific chemical, e.g. catalase, to a nutrient medium*

* Exposure of yeast extract media, particularly if it has been autoclaved, to light can cause the production of superoxide radicals and hydrogen peroxide—see p. 35 (Hoffman P.S., Pine L. & Bell S. (1983) *Appl. Envir. Microbiology* **45**, 784–791). Substances other than catalase, e.g. 3.3′-thiodipropionic acid, have been recommended (McDonald *et al.* (1983) *Appl. Envir. Microbiology* **45**, 360–365.

physiologically damaged but not killed by preservatives, processing methods or long-term storage under adverse conditions such as those imposed by an intermediate moisture food (Table 3.1). There would be little prospect of such organisms growing if they were placed directly into a selective medium. It is for this reason that in the isolation of *Salmonella* spp., for example, a resuscitation step—incubation of the food sample with peptone water or some other non-selective medium—is taken before incubation of the material in Selenite or Tetrathionate broth. The actual choice of media and methods for routine use will be influenced by the results of comparative surveys, such as those sponsored by the International Commission on Microbiological Specifications for Foods (1978).

SPECIFICATIONS

The sampling and analysis of a finished product is only one stage in the routine monitoring of foods. The ingredients used to make a food and the methods of manufacture should be considered also for how else can a sensible specification be arrived at? What are specifications and how are they formulated? They may be decided upon by some regulatory agency of a government, by a retailer who seeks assurance that the suppliers of foods are reaching consistently satisfactory standards, or the manufacturer who needs to assess the performance of his factory. It will be appreciated that this is an important aspect of food microbiology. It is, however, one which is the cause of much argument and controversy and one which is bedevilled with semantics, hence the terms microbiological standards, specifications, guidelines or bacteriological reference values. Suffice it to say that if a specification is adopted, it should be flexible and reflect an inherent feature of quantitative microbiology, the relative inaccuracy of all methods used for viable counts. To emphasize this flexibility, the following discussion will consider three approaches to the establishment of standards.

Within a factory the following categories might be adopted for the day-to-day monitoring of a quality assurance programme.

Target. That level of microbial contamination which assures the safety of the consumer, the well-being of the food during storage and distribution, and which is achieved when *every* phase of the manufacturing process is working at maximum efficiency. With canned foods having the potential to allow the growth of *Clostridium botulinum*, the target standard is the only acceptable standard. With many foods, the target provides the ultimate goal *under* the existing methods of production—to change the methods could well impose costs which would be prohibitive and which need not provide any additional benefit to the consumer.

Acceptable. That level of microbial contamination which assures the safety of the consumer and the well-being of the product and which is achieved under normal (and by inference, acceptable) conditions.

Suspicious. That level of microbial contamination which indicates some failure or fault in the manufacturing process but, from experience, is not considered to impose a threat to the consumer. If it is a perishable food, the predicted incidence of spoilage will have to be considered also.

The actual numbers and types of organisms used to establish these categories will be determined by the type of product and the storage conditions obtaining during distribution and storage in the shop and home. In practice, therefore, the media and methods ought to provide information on the numbers of organisms that are physiologically equipped to cause spoilage under the envisaged conditions of storage. The reputation of the product as a vehicle for food-poisoning organisms or their toxins will dictate the choice of media and methods also (see Chapter 9). There has been a tendency for food microbiologists to adopt the precept of their colleagues in water microbiology and to consider the use of indicator organisms as an index of the probable incidence of contamination of foods with *Salmonella* spp. In water microbiology (see p. 145) the value of *Escherichia coli* as an indicator organism has been established by long experience. As very few *Salmonella* spp. ferment lactose, a search for glucose (Enterobacteriaceae) rather than lactose (coliforms) fermenting organisms has been adopted for examination of foods for indicator organisms. Although experience has led some investigators to extol the utility of indicator organisms in the routine analysis of processed foods, others, for example the International Commission on Specifications for Foods, have little confidence in the reliability of such an approach.

When devising categories such as those noted above, the specifications must cover bands rather than select one statistic for each category. With the latter approach there is the danger of a 'golden number' being imposed and endless argument between those who analyse and those who administer; for example, what significance would you attach to analyses which revealed counts of 1.1×10^3 or 9.8×10^2 organisms g^{-1} if the specification stated boldly that less than 1.0×10^3 was the standard for a certain product?

If guidelines such as these are adopted, then some tolerance level has to be established, say in the analysis of 10 samples, not more than 2 should be in the suspicious category.

CONTROL METHODS

The method derived by Mossel to establish microbiological reference values for foods takes account of the need for critical selection of media and methods for the isolation of appropriate organisms as well as the need to accommodate the scatter of results obtained during a survey of 10 randomly selected samples from each of 10 or more factories in which prior inspection had established that Good Manufacturing Practices were observed. A skewed distribution is a common feature of the results from such a survey (Fig. 3.1). The curve permits the selection of the count which is not exceeded by 95% of the samples (ϕ). This figure may be one to several log cycles away from N, the maximum count that would be expected in products from a well-managed factory. The selection of values for N will be dictated by knowledge

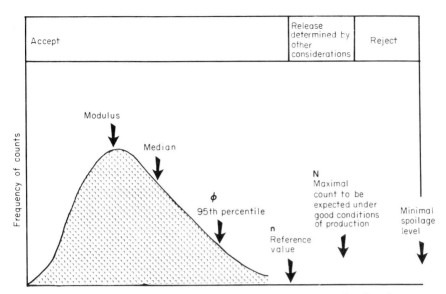

Fig. 3.1. The distribution of results obtained from a survey of several samples of products from several factories in all of which good manufacturing practices are observed. For further details, see text (redrawn from Mossel D.A.A. (1976) in *Intermediate Moisture Foods* (Eds Davies R., Birch G.G. & Parker K.J.) Applied Science Publishers, London).

about the potential risk to the health of the consumer or spoilage of the product. Should ϕ be close to N, then manufacturing practices need to be examined so that improvements in the microbiological quality of a product can be achieved. When ϕ is far removed from N, then a reference value (n) can be selected. With spoilage organisms, a tolerance has to be adopted in the selection of n so that variations in the distribution of contaminants and accuracy of microbiological methods are accommodated. Likewise a decision has to be made on the number of samples that will be accepted in the range n–N. When considered alongside the standards discussed initially, it is evident that this latter approach gives three categories also: acceptable $(0-\phi)$, suspicious (n–N), and unacceptable (N+).

A third example of the formulation of specifications is provided by ICMSF. The basic concept relates to the type and stringency of a sampling plan, the hazard potential of a food, and the conditions in which a food is expected in the usual course of events to be handled and consumed after sampling (Table 3.2). With this approach a food is assigned to a category (Case) which associates the health hazard (*none, low and indirect; moderate and direct, with limited or potentially extensive spread; or severe and direct*) with the expected conditions of handling of the food (*conditions reduce, cause no change or increase concern*). Thus the plan accommodates the anticipated extremes, ranging from Case 1 (*no health hazard and low incidence of spoilage*) to Case 15 (*health hazard severe and direct and possibly influenced by conditions of handling*). The stringency of sampling reflects this situation; with Case 1, five sample units (n) with three defective samples (c) is acceptable, whereas c = 0 with n = 60 in Case 15.

Table 3.2. Suggested sampling plans for combination of degrees of health hazard and conditions of use ('cases')*. (Reproduced, with permission, from Clark D.S. (1978). The International Commission on Microbiological Specifications for Foods. *Food Tech. (Overview)* 32 *(January)*, 53.

Degree of concern relative to utility and health hazard	Conditions in which food is expected to be handled and consumed after sampling, in the usual course of events†		
	Conditions reduce degree of concern	Conditions cause no change in concern	Conditions may increase concern
No direct health hazard			
Utility, e.g. shelf-life and spoilage	Case 1 3-class, n = 5, c = 3	Case 2 3-class, n = 5, c = 2	Case 3 3-class, n = 5, c = 1
Health hazard			
Low, indirect (indicator)	Case 4 3-class, n = 5, c = 3	Case 5 3-class, n = 5, c = 2	Case 6 3-class, n = 5, c = 1
Moderate, direct, limited spread	Case 7 3-class, n = 5, c = 2	Case 8 3-class, n = 5, c = 1	Case 9 3-class, n = 10, c = 1
Moderate, direct, potentially extensive spread	Case 10 2-class, n = 5, c = 0	Case 11 2-class, n = 10, c = 0	Case 12 2-class, n = 20, c = 0
Severe, direct	Case 13 2-class, n = 15, c = 0	Case 14 2-class, n = 30, c = 0	Case 15 2-class, n = 60, c = 0

* From ICNSF (1974); used with permission of the University of Toronto Press and ICMSF.
† More stringent sampling plans would generally be used for sensitive foods destined for susceptible populations.

DISPERSAL OF MICROORGANISMS

There are four main approaches to the control of microbial contamination or the dispersal of microorganisms:
1 hygiene and quality control;
2 removal of organisms;
3 asepsis;
4 deliberate infection with selected organisms (Chapter 5).

Only the approaches dealt with in (1) and (2) will be considered in this chapter.

(1) Hygiene and quality control

In this category, the food microbiologist working along with *all* members of the team responsible for producing a food has to consider two major means of dispersal of microorganisms:
(A) inanimate carriers;
(B) special sources of animal infection.

(A) INANIMATE CARRIERS

As an aid to the identification of important sources of contamination in the processing of a food, the following should be considered:
  (i) air;
 (ii) water;
(iii) ingredients;
 (iv) equipment;
  (v) processing;
 (vi) spoiled foods.

(i) *Air*

From a practical viewpoint, the importance of air as a vector has to be considered in terms of the organisms it carries and the extent to which it makes contact with a product. Thus when large volumes of air are needed for the dehydration of foods, the numbers of airborne organisms may need to be reduced by filtration. Likewise in the aseptic packaging of a processed food, the potential for air to cause recontamination must be considered.

The meat industry provides an interesting example of infection arising from dispersal of moulds by air. Meat pies need to be cooled rapidly after cooking so that there is minimum disturbance in the process of making, baking and wrapping in transparent film. Cooling can be achieved simply by placing pies on racks in rooms having a rapid exchange of chilled air. To ensure turbulence, the walls are shuttered and the air is forced between the slats of the shutters. Crumbs can collect behind the shutters and, if good housekeeping is not practised, moulds can grow and discharge spores. If these enter the holes in the lid of pastry, a colony can develop; in the early stages, growth will not be observed until the pie is opened. If the spores alight on the surface of the lid, they do not germinate because the glazed surface seems to provide a barrier to the diffusion of water—and probably of antigerminants if these be in the mould spores. Mould growth on the surface can be a problem if warm pies are wrapped and then chilled. Water condensing on the wrapping film drips on to the surface of the pastry lid and moulds can grow on the moist pastry. In addition to sensible management and good housekeeping, monitoring of the production area and chilling rooms by the 'settle plate technique'—a Petri dish containing malt agar is opened in the production areas—provides the microbiologist with the means of assessing the efficacy of any policy of control. As mould spoilage tends to be a feature of the warmer months in the northern hemisphere, any control policy ought to reflect seasonal influences.

In the propagation of 'starter' cultures for the dairy industry, it is essential to prevent contamination of the bacteria with phage. The use of a water seal over the injection port (Fig. 3.2) provides an example of a simple means of minimizing contamination.

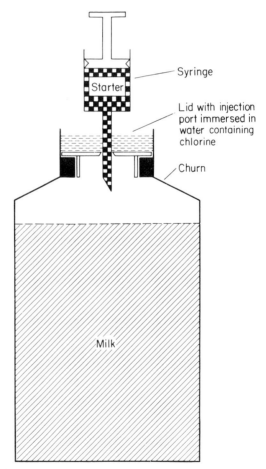

Fig. 3.2. A method adopted by the dairy industry to protect starter cultures from phage infection when subcultures are made.

(ii) *Water*

Through its ubiquitous use in food processing, water can be considered as an ingredient, an important agent in the dispersal of microorganisms, or, when used as ice, the means of achieving chill storage of foods under conditions, such as those obtaining on small fishing boats, where mechanical refrigeration is not possible. In addition it is widely used to wash and thereby reduce the level of contamination (examples given in (iii)) of foods, and as a medium for exchanging heat with a food, either taking heat to a food, as with the pasteurization of milk in a heat exchanger, or taking away heat as in the coolers used in the canning industry. In these applications, water will be a part of the food or it will make contact, either intentionally or accidentally, with a food. Thus it is essential that it be of a satisfactory microbiological standard; it must be free of microorganisms of faecal origin and contain few organisms capable of growth at low temperatures (Chapter 7).

The processing of broilers provides an example of a situation in which water can play a part in the selection and dispersal of microorganisms. After plucking and evisceration, the carcass is cooled by immersion in water containing ice. As a large number of carcasses move along a cooling tank in the course of a day, it is obvious that water can effect the transfer of organisms from one carcass to another, providing of course that the turbulence of the water is sufficient to remove organisms that may be adhering to the skin. Uncritical management of this stage of broiler processing could lead to an increase in the number of psychrotrophic micro-organisms present in the cooling tank and rapid spoilage of carcasses held at chill temperatures. Problems associated with the chill tank can be ameliorated by plant design (counterflow of chickens and water is to be preferred; Fig. 3.3), the use of 2.5–6.0 litres of water per bird processed so that contaminants are washed out of the system, and chlorination (up to 200 p/10^6) of the water.

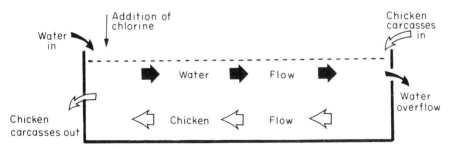

Fig. 3.3. In the chilling of poultry with slush ice, a counterflow of carcasses and water is one contribution to the control of microbial contamination.

(iii) *Ingredients*

Sensible policies for controlling the dispersal of microorganisms by ingredients cannot be arrived at until there has been a thorough assessment of the contribution, not only of *numbers* but also physiological types of microorganisms, which an ingredient will make to a *particular* food; for example, spices can be the most heavily contaminated ingredient of British fresh sausages (up to 10^7 organisms g^{-1}) but, through being a minor ingredient, the contaminants are diluted by the other ingredients. Moreover, the contaminants of the spices, mainly aerobic spore-forming bacteria, do not make a contribution to the microbial association of this type of sausage which is not cooked until required for a meal. With a related food commodity, a liver sausage, the heat treatment used during manufacture will kill vegetative bacteria and the spore-formers can then germinate and form the microbial association. If this type of analysis identifies an ingredient such as a spice to be an important source of contaminants, then a method of control has to be sought. With spices, ethylene oxide has been used to reduce the level of contamination or extracted oleoresins used instead.

Examples of ingredients which may contribute to contamination of processed foods are given in Table 3.3. With flour or gelatine used as

Table 3.3. Some ingredients which contribute contaminants to processed foods.

Ingredient	Food
Spices	Soups, meat products
Flour	Soups, bread, canned foods
Solar salt	Salt fish
Syrups	Jams, confectionery
Gelatine	Soups, meat products

'thickening' agents, the spores of many species of *Bacillus* and *Clostridium* may be added to a food. The fate of the contaminants will be determined by the nature of the food and the processing methods; for example, if salt-tolerant denitrifying *Bacillus* spp. were present in ingredients used in the canning of hams or bacon, then their growth by means of anaerobic respiration with nitrate as terminal electron acceptor (Fig. 8.15) could result in the formation of dinitrogen and gaseous oxides of nitrogen. These would cause a bulge in the ends of the cans. Thermophilic spore formers present on the beard and in the crease of wheat grains can contaminate flour and achieve numerical dominance if the flour is steamed, as is often the case with that used for 'thickening' soups or the gravy in meat products. In this example, the grain and flour are acting merely as vehicles for the dispersal of microorganisms of soil origin. Similarly, soiling of vegetables can introduce thermophilic spore formers to canned foods. This can be overcome in part by thorough washing of the vegetables; an example of water being used to remove microorganisms.

In the examples given above, an ingredient acts as a vehicle for the contamination of a product with microorganisms which may cause spoilage if a process is not controlled or storage conditions are unsuitable. There are situations where an ingredient contributing to preservation can initiate spoilage because it contains organisms which are physiologically adapted to grow on the preserved product. Thus the use of solar salt containing halophilic bacteria can cause spoilage of salt fish. Osmophilic yeasts or xerophytic fungi in concentrated syrups could cause spoilage of foods whose preservation depended in part on the binding of water by a soluble carbohydrate. The syrups may contain yeasts which are not osmophilic and these will grow rapidly once the syrup is diluted. Processing methods can influence markedly the response of osmotolerant yeasts to a new environment. Thus rapid growth can occur if the latter has an a_w comparable to the syrup but there can be a marked decline in the population of yeasts if a syrup is diluted before mixing with a material having a low a_w, this effect has been ascribed to osmotic shock. Extensive and rapid growth may follow the death phase. The dilution of a syrup may also lead to the growth of non-osmotolerant yeasts that had been quiescent during the storage of the product.

Standards for ingredients. When the relative importance of an ingredient as a vector has been determined, then an attempt can be made to define standards for microbial contamination which are appropriate to the food and process. There are at least three approaches to their implementation.

1   Contracts with suppliers to supply ingredients which comply with the standards. This is a practical approach with non-perishable products such as flour and spices, for example, because the material can be analysed before use.

2   Use of 'sterilized' material or, in the case of spices, those treated with ethylene oxide.

3   Removal of organisms before the raw material is used (Table 3.4).

In some countries, trade associations have established standards. Thus in the USA, the National Canners Association have set standards for starch and sugar that are intended for use in canning. As would be expected, they emphasize the levels of contamination of these two ingredients with thermophilic spore formers. Likewise, the American Bottlers of Carbonated Beverages have established standards for granulated and liquid sugar, the emphasis being laid on levels of contamination with yeasts and moulds.

Microbial products. At the beginning of this section, emphasis was given to the need to assess the relative contribution, and importance, of an ingredient to the overall microbial contamination of a product. Similarly there is a need to assess the relative importance of contaminants or their products present at the end of a process. Thus with the bulk storage of cider, the growth of acetic acid bacteria on the surface of the cider may be commercially acceptable because of the prohibitive costs associated with the removal or inhibition of the spoilage organisms and the probability that, because of relative volumes, the dilution of acetic acid would be such that unacceptable changes in flavour of the product would be unlikely to occur. An opposite view would have to be taken with small volumes of cider where appearance and flavour would be changed by the growth of acetic acid

Table 3.4. Methods used to reduce the level of contamination of foods.

Method	Applied to
Spraying with water (efficacy improved by adding acetic acid)	Carcasses of beef and pork
Immersion in boiling water	Pork used in the preparation of ham and bacon
Washing with water	Vegetables, especially those used in canning where thermophilic spore formers can cause spoilage
Addition of chlorine to water	Control of contamination of broilers during chilling
Sparging of chlorinated wash water	Preparation of oysters
'Blanching'—dipping in hot water	Vegetables, especially those used in canning
Pasteurization	Milk used for butter and cheese
Sulphiting	Fruit juices used in cider and wine production
UV irradiation	Meat in cold stores
Storage in 'clean' sea-water and spraying with sea-water—chlorine may be added	Oysters and other shell fish

bacteria. Pasteurization, filtration or enriching the head space with CO_2 would have to be considered as possible means of preventing spoilage.

In addition to the contamination of an ingredient with the viable cells and spores of microorganisms, attention might have to be turned to changes resulting from microbial products even though the organisms that produced them are dead or quiescent. An example has been discussed previously, the production by acetic acid bacteria growing on apples of metabolites which react with SO_2. Other examples are provided by the breakdown of the pectin bridges between cells of plant material or fruits which are being stored in bulk prior to manufacture or packing into small containers. Some organisms associated with the maceration of plant materials are given in Table 5.3.

The maceration of sulphited cherries (used in the production of maraschino or glacé cherries) has been associated with the action of polygalacturonases formed by *Cytospora leucostoma* infecting fruits on the trees. Although infection may be relatively slight, the macerating enzyme is extracted into the sulphited liquor with the result that all the fruit are bathed in a dilute solution of enzymes. A similar problem is encountered by those who store sulphited strawberries for jam production. If the fruit in the field is infected with *Rhizopus stolonifer, Rh. sexualis* or *Mucor piriformis* the macerating enzymes, polygalacturonases, are released from infected fruit into the sulphited liquor and, although they are inactivated slowly, the fruit can be reduced to a purée. Fruit infected with *Botrytis cinerea* may not break down in storage because this organism's pectinase is denatured quickly in sulphited liquor; this pathogen can also have a sparing action on fruit infected with the organisms noted above. With this example, the food microbiologist has to consider the control of microorganisms during the *production* of the crop and to establish methods of handling and storage which will minimize the action of any enzyme which may be brought into the factory. The following recommendations were made to jam manu-facturers in the UK:

1 ensure that a spraying programme minimizes infection of the fruit (strawberries);

2 reduce to a minimum the period between picking and packaging—use chilled storage if delay is unavoidable;

3 pick strawberries when they have just ripened—overripe fruits are easily damaged;

4 eliminate infected fruit—as little as 1% infection of fruit can lead to maceration—and do not use those end-of-the-season fruits which are unfit for the retail market;

5 use clean utensils and barrels lined with plastic bags; ensure that the SO_2 is distributed evenly during the packing of barrels with fruit.

(iv) *Equipment*

When considering the role of equipment in the dispersal of microorganisms, two important factors have to be recognized:

(a) the levels of microbial contamination of ingredients at the start of production at the beginning of the day—it being assumed that the equipment is clean;

(b) the design and layout of the equipment.

It is probable that an equilibrium will be established on a production line between the level of contamination of a product and that of the equipment—and possibly the hands of the workers. Control in this case must be directed at the microbial load present on raw materials and those extrinsic factors which hinder the growth of organisms transferred to equipment. In Chapter 1 the control of flow rate was given as an example of a means whereby the growth of thermophiles in a hot, nutrient liquid was controlled. There are probably other situations where this concept can be applied. In the butchering of meat, the control of ambient temperature (40–50 °F) may well provide a means of slowing the growth of micro-organisms which are retained on equipment and thereby maintaining a situation where the equipment is transferring back to the meat only a few organisms even after several hours of production.

In the design and installation of equipment, there is the need not only for a piece of equipment to do a specific job but also for it to be designed, manufactured and installed so that it can be cleaned easily and thoroughly by semi-skilled workers without the assistance of engineers. It is imperative that the following faults are eliminated at the design stage:

1 'dead end' pipes;
2 the joining of metal pipes with flexible tubes and Terry clips;
3 open joints;
4 gaskets which through hardening become brittle and crack;

It is also important that the avoidance of the following potential faults is emphasized in the instructions to the manufacturers:

(a) rough welds;
(b) the use of metals which, if in contact, will form cells when an electrolyte is present.

The efficiency of disinfectants is decreased when they have to compete with organic material. Moreover it is rarely possible to chemically sterilize nooks and crannies unless the disinfectant has a *very low surface* tension, particularly if they contain organic debris. It is considerations such as these which should guide the microbiologist who has the opportunity of advising on the design of equipment for the food industry.

In many cases, the food microbiologist has to formulate a policy for controlling the dispersal of microorganisms in a factory where the equipment was designed with little attention having been given to the methods of cleaning, and its installation was done with little concern for the needs of the people who have to do the cleaning. In such a situation, a distinction has to be drawn between good housekeeping, aesthetic cleaning, and cleaning which will rid the equipment of organisms adapted to spoil the food being processed. The former is concerned largely with those bits of the machine and its surrounds with which food should not make contact. With the latter, a cleaning schedule—the detailed instructions to the cleaners—can only be prepared after a detailed microbiological analysis has established those parts or pieces of equipment which need special treatment if they are to be freed of spoilage organisms. In such an analysis, the general principles of enrichment culture should guide the search for the crevice in which the *particular* spoilage organism is maintained with inappropriate cleaning from one day to the next. The same analytical approach may be needed to locate the niches of infection in equipment which, although designed for a specific function and with the problems of cleaning in mind, causes contamination of a product with a spoilage organism. Thus problems can be encountered with a closed system design for the bottling of a yeast-free acid liquid and for 'in-plant' cleaning at the end of production. If the pressure used during bottling is greater than that used during cleaning, and if the pressure sensors are of the piston and cylinder type, then it is possible for yeast to colonize that part of the cylinder which is not exposed during cleaning but is during bottling.

When the methods for efficient cleaning have been established there is the requirement to ensure that they remain effective in the day-to-day operation of a processing unit. This can be done by checking randomly the extent of contamination of cleaned equipment. Several methods are available for this:

I   The swab–rinse technique: a sterile cotton or alginate swab is moistened with sterile Ringers solution or other appropriate medium. It is rubbed over a known area; the swab is shaken or, with alginate, dissolved in diluent and known volumes of the diluent plated on an appropriate medium.

2   Rinse technique: surfaces—especially those within pipes—are rinsed with a sterile diluent, such as Ringers solution, some of the diluent is collected and plated on an appropriate medium.

3   Agar contact method: the sterile face of a block of nutrient agar is pressed against the surface to be examined and then incubated.

4   Direct surface planning: a sterile, molten nutrient agar is poured on to the surface to be examined. When the agar has gelled, it is protected from aerial contamination by a sterile cover and, after incubation, the number of colonies forming at the interface of the surface and agar are counted.

(v) *Processing*

In the preceding section, the discussion was limited to particular facets of the problems associated with the dispersal of microorganisms. In practice, of course, many of these may contribute to the contamination of a product, particularly if there are several stages or phases in the preparation. It is probably essential to do a thorough microbiological analysis of a process so that a complete picture is obtained. This is sometimes referred to as 'line analysis' when the influence of processing on the contamination of a product with spoilage organisms is being considered. The term, longitudinal integrity of processing for safety, has been applied to analyses concerned with the elimination or control of pathogens. A hypothetical example of the former is given in Fig. 3.4. This illustrates a fault with two potential causes: (a) heavy contamination of a product from ingredients or working surfaces, or (b) heavy contamination of the finished product caused by failure to control the temperature at the various stages of manufacture.

(vi) *Spoiled foods*

It is perhaps a statement of the obvious that spoiled foods should not make contact with fresh foods or ingredients. However, it is the practice of some manufacturers to collect daily the products which have been the subject of complaint by retailers or the consumer, or those products which have not been sold within an accepted storage life. Such material will provide the food microbiologist with a means of assessing the efficacy of those standards and policies which have been introduced for assuring the shelf-life of a product. Moreover, it may provide an early warning of a fault, perhaps an unusual one, which has not been demonstrated by the routine methods of control. The returned material will also provide the inexperienced microbiologist with many examples of the exceptional insults to which foods are exposed during distribution and storage. Having recognized the potential value of this daily inspection, provision must be made for it to take place

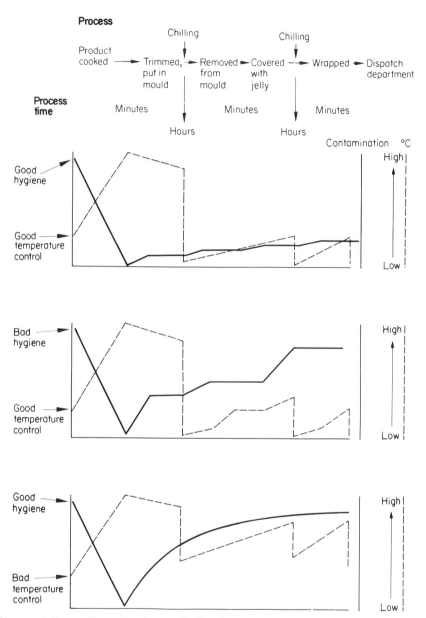

Fig. 3.4. A diagram illustrating the contribution of good temperature control and good control of hygiene to microbiological quality of a product.

in an area remote from the production areas and from the dispatch department. One is dealing with enriched cultures of organisms which are physiologically equipped to spoil a particular food and these must not be allowed to seed a fresh product.

In certain manufacturing processes, the remnants of materials from certain operations are collected for re-use. Thus in the bakery industry fermentations due to osmophilic yeasts can cause problems with fondant that is used to coat cakes. The yeasts do not grow in fondant at the operating temperature of *c*. 40 °C but they can do so at ambient temperatures in scraps kept overnight or over a weekend. Likewise, yeasts can cause fermentation of marzipan coatings on cakes, again due to build up in their numbers on scraps from previous processing. Indeed yeast growth in the scraps can be accentuated due to an increase in a_w resulting from contamination of the marzipan with cake crumbs. Another example of problems arising from the use of scraps is provided by the meat industry. In the manufacture of meat pies, circular or oval shapes are cut out from rectangles of pastry or the base of the pie is pressed out in a dye and the excess pastry trimmed off. The scraps, sometimes contaminated with meat juice, are re-worked or blended with new batches of pastry. Uncritical supervision of the re-use of dough or its storage overnight or over a weekend can lead to fermentation and tainting of the pastry. This can be a serious problem with uncooked pies held at chill temperatures because the contaminants, mainly yeasts, continue to grow albeit slowly; in extreme situations the level of contamination of the pastry, particularly during the summer months, can impart a taint even when the pies are cooked immediately following preparation. As the pastry becomes acid due to fermentation, the monitoring of pH may provide a control system for management.

(B) SPECIAL SOURCES OF ANIMAL INFECTION

Man should be regarded as the most important 'animal' vector of microorganisms in a food factory, particularly if movement within the factory is uncontrolled. Thus in the example discussed above, the isolation of spoiled foods from the production area will be of no avail if persons who inspect the spoiled food carry organisms on their hands and clothing to those areas where foods are being prepared. To minimize the extent to which man acts as a vector, the following should be considered:

(a) instructions in hygiene and elementary food microbiology;
(b) provision of a clean working environment having clean rest rooms and adequate toilet facilities;
(c) provision of well-designed equipment—an attempt to prevent the dispersal of microorganisms via the hands from equipment to a food;
(d) protective clothing—water-repellent so that microorganisms cannot colonize sodden fabrics.

Rats, mice and flies are often considered to be vectors of microorganisms, particularly pathogens. The design and management of the factory should be such that these vectors cannot enter, let alone colonize, the factory. Insects have been implicated in the dispersal of osmophilic yeasts. Thus the infection of honey can have its genesis in the infection of the flower from which the nectar was taken. A recent study indicated that acetic acid bacteria may overwinter in bee hives also. There is evidence also that the fruit fly, *Drosophila melanogaster*, plays an important role in the dissemination of yeasts (*Hanseniaspora uvarum* and *Pichia kluyveri*) which ferment tomatoes, particularly those fruits with broken skins. Fruit flies have been implicated also in the transmission of yeasts which bring about spoilage of Calimyrna figs in California.

(2) Removal of organisms

Figure 2.12 illustrates the relationship between initial level of infection and the storage life of meat. It is obvious that a long shelf-life cannot be achieved unless the contribution of *all* sources of contamination—equipment, the hands and clothing of the workers, etc.—are minimized or the contaminants removed. The methods which have been used in attempts to reduce the levels of contamination are given in Table 3.4.

The treatment of certain molluscan shell fish provides an example of a situation where water is the vector of contaminants and then, under controlled conditions, it is used to rid the fish of contaminants. If shell fish live in polluted water, then they can be contaminated with faecal organisms, coliforms and salmonellae, and these organisms may be concentrated by the feeding methods of the shell fish. The method of filter-feeding results in microorganisms being trapped on the gills and then taken by cilia action to the mouth. If transferred from polluted to non-polluted water, mussels will be freed of coliforms within 48 hours—the microorganisms through being trapped in the mucous thread of the faeces are prevented from contaminating the bulk of water. Thus a simple method of husbandry, relaying in non-polluted water, can be used to rid the shell fish of coliforms. If non-polluted water is not available, then the shell fish can be held in tanks and supplied with sea-water in which the coliforms have been killed. In one method, the mussels are stored in tanks to which is added treated sea-water—sea-water containing 3 p/10^6 of chlorine is stored for 8 hours and then the chlorine neutralized with thiosulphate. After 24 hours, the tank is drained and a jet of water used to flush out debris. The mussels are stored in treated sea-water for a further 24 hours and then for one hour in chlorinated sea-water. In another method, the sea-water is circulated through the shell fish and then it is pumped to the top of a small tower where it passes as a thin film beneath a UV lamp before cascading down the tower and into the tank. Thus coliforms are killed and the water aerated as a continuous process.

4

Pasteurization, Appertization, Radiation and Asepsis

With the heat treatment of food, two quite separate objectives have to be recognized. With cooking, for example, the process is intended primarily to improve the palatability of a food. The attention that needs to be given to recontamination (or asepsis) will be dictated by the intended use of the food after cooking. Little attention needs to be given if the food is to be consumed immediately; stringent temperature control and aseptic or clean handling are called for if the food is to be served cold, especially if further preparative stages, such as deboning of cooked chicken, are done without additional heat-treatment (Fig. 9.11). An account of the microbiological problems associated with cooked foods is given in Chapter 9. The term blanching is applied to processes in which vegetables are treated at 100 °C for 2–5 minutes; the objective may be to inactivate enzymes that cause changes in colour, flavour and nutritive properties during the storage of frozen or dehydrated products, or to remove gases, inactive enzymes and aid the packing of vegetables in cans. Of course many vegetative micro-organisms will be killed, but this benefit will be secondary to those of enzyme inactivation, etc. In other applications of heat to foods, the objectives are (1) to kill pathogens in the product, or (2) to kill those organisms that cause spoilage. Three temperature ranges are used to achieve one or both of these objectives: 70–100 °C, temperatures around 121 °C and temperatures of 132 °C or above. The term pasteurization is applied to the first, retorting to the second and High Temperature Short Time (HTST; see p. 4) the third. The general term, appertization, is used for processes involving the last two temperature ranges. The success of all these processes will depend upon the processed food being protected from recontamination (asepsis).

PASTEURIZATION

The destruction of peroxidase and catalase, commonly the most heat-resistant enzymes of plant tissues, is used to establish and then to monitor a blanching process, the rationale being that a process that inactivates these two will ensure the organoleptic stability of the product. The choice of one or other of the three temperature ranges given above will be influenced by considerations of the purpose of the process and the chemical and physical composition of the food. Thus with milk, pasteurization (62.8 °C for 30 min) was used initially in the UK to break the infection route of *Mycobacterium tuberculosis*; only subsequently did the improved storage history of the processed milk, especially with chill storage, become an important commercial consideration. Likewise the pasteurization of whole eggs (a mixture of yolk and white) at 64.4 °C for 2–5 minutes is intended to control the dissemination of *Salmonella* spp. Indeed this time–temperature combination was chosen as a result of trials in which the most heat-resistant member of this genus, *Salmonella senftenberg*, was used and the extent of damage of the functional properties of the egg assessed so that a high probability of freedom from infection could be achieved without unacceptable damage to the product. Although pasteurizing equipment is fitted with charts for recording times and temperatures, a chemical test for adequate processing is essential. Indeed such a test may indicate that a pasteurized product has been contaminated with untreated material. With milk, a phosphatase, and with whole egg, an α amylase, have been selected for such tests. In other applications of pasteurization, for example with cider and beer, most if not all organisms capable of causing spoilage are killed although the time–temperature combination would not be expected to sterilize the product. With sweetened condensed milk, a mild heat treatment is acceptable because of the moderate heat resistance of the osmophilic and osmotolerant yeasts, the only organisms that could grow at the a_w imposed by the addition of sucrose. Similarly a mild heat treatment of canned ham has proved commercially acceptable, providing the product is stored at chill temperatures, because spore formers fail to grow due to the pH and the preservatives, NaCl and NO_2^- (see Chapter 9 for further details).

APPERTIZATION

Of the many factors which determine the time–temperature combinations to be used for retorting or HTST treatment, the heat resistance of bacterial endospores, enzymes and the nutrients of the food is of cardinal importance. It must be recognized also that a particular time–temperature combination may well produce a shelf-stable product without freeing it of viable endospores. The latter must of course be of non-pathogenic species; they must also be of species that are unable to grow in the environment present in the processed food. It is for this reason that such unfortunate terms as 'commercially sterile' have been applied to heat-processed foods; appertized would be a preferable term because it removes the need to qualify sterility.

When appertization is being considered, there is a need to assess the contribution of the following:
(i)   the inherent properties of the contaminating organisms;
(ii) the environmental factors obtaining at the time the heat is applied;
(iii) the environmental factors obtaining during the storage of the processed food.

The inherent properties of microorganisms

An introductory course in microbiology covers general features of the resistance of microorganisms to heat. Thus vegetative cells are killed rapidly by heat whereas resting stages, particularly the endospores of *Bacillus* and *Clostridium* spp., can be very resistant. Moreover, the extent of an organism's resistance may be modified by age (bacterial cells tend to be most resistant to heat in the stationary phase of growth), and the temperature obtaining during the incubation of the culture, increasing the incubation temperature of a culture can increase the thermal resistance of endospores. The vegetative cells of some dairy microorganisms are remarkably tolerant to heat, for example *Streptococcus faecalis* (withstands 60 °C for 30 min) and *Microbacterium lacticum* (the original isolates tolerated 72 °C for 15 min). For the control of processes in the canning industry, a precise index of an organism's or resting stage's resistance to heat is required. This may be established by determining the rate of reduction of viability of the test organism in a medium to which heat has been applied for known times. With this method, several samples of the suspending medium will be analysed and when the results are plotted ($\log_{10}$ number of surviving organisms against time in minutes), a straight line will be obtained (Fig. 4.1). The relationship

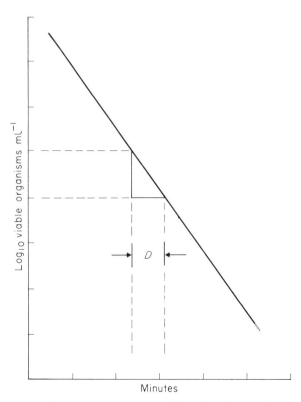

Fig. 4.1. Survivor curve for microorganisms suspended in a liquid and subjected to a constant, lethal temperature.

between time and the numbers of survivors in a population subjected to a known temperature can be stated thus:

$$(\log_{10} n_0 - \log_{10} n) \times D = t_n$$

where n_0 = the initial number of spores, and n = the number after t_n minutes.

This formula provides an index of an organism's resistance to heat, the D value—the number of minutes of heat required to cause a reduction of the number of viable organisms by a factor of 10. This value permits also a comparison of the heat resistance of different organisms, particularly their endospores, viz. (temperature, 100 °C): *Bacillus stearothermophilus* (3000 min), *Clostridium botulinum* A and B (50 min), *Cl. perfringens* (20 min), and *Cl. pasteurianum* (0.5 min).

As the death rate of spores is logarithmic—a constant proportion of viable endospores is killed in unit time at a selected temperature—then in practice it has to be appreciated that the probability of a food being freed of viable endospores with a heat treatment will be a function of time and the initial level of contamination. Thus a 5 D treatment of 10^6 spores ml^{-1} will leave a viable population of 10 ml^{-1}; the same treatment of a 10^3 suspension will leave 10^{-2} ml^{-1}. If, therefore, 100 samples each containing 10^3 spores were processed, then probability would dictate that the latter treatment could yield one sample containing viable endospores. It is this line of reasoning which led to the adoption of the 'botulinum cook' or 12 D concept, that heat treatment which will reduce the viability of a population of spores of *Clostridium botulinum* by 12-log cycles. As the initial level of infection of foods with this organism is low, then one can think in terms of the number of treated samples (cans) rather than an individual can. The very rare occurrence of botulism due to the consumption of canned foods bears witness to the usefulness of this concept.

It will be recognized, of course, that the D value refers only to a temperature and it is stressed in discussions of sterilization by heat of bacteriological media, glassware, etc., that time and temperature are inversely related, sterilization at a high temperature is achieved more quickly than at low temperatures. It has been shown that the D value varies with temperature (Fig. 4.2), viz.:

$$\frac{t_1 - t_2}{z} = \log D_2 - \log D_1$$

where D_1 = the D value at temperature t_1 and D_2 = the D value at temperature t_2. The z value, a measure of the temperature coefficient, is that change in temperature which changes the D value by a factor of 10 and thus permits a comparison of the influence of temperature in time–temperature combinations. So far the methods have not provided a 'datum point', a standard temperature to which treatments at other temperatures can be referred. A temperature of 121 °C has been widely adopted and the F value defined as the number of minutes of heating which produces the same killing of bacterial spores as the process which requires to be specified. If

72

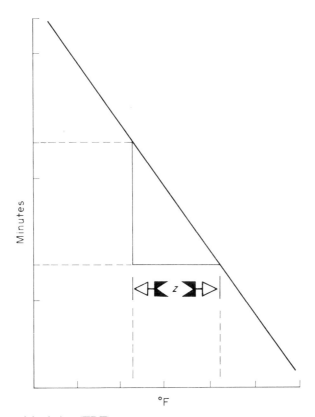

Fig. 4.2. Thermal death time (TDT) curve.

provided with values for D and z, a 'thermal death time' (F) can be derived.

The above discussion is brought into focus by a consideration of the heat treatment that needs to be given to canned foods of pH greater than 4.6 so that there is a very high probability of freedom from viable spores of *Clostridium botulinum*. On the assumption that this organism has a D value at $121\,°C$ of 0.21 min and that each can of food contains one spore of this organism, then a process of $121\,°C$ for 2.52 min would reduce the population density to one spore in 10^{12} cans, viz.:

$$F_0 = D_{250}(\log_{10} x - \log_{10} y)$$
$$= 0.21(\log 1 - \log 10^{-12})$$
$$= 0.21 \times 12 = 2.52.$$

For additional safety, $F_0 = 2.52$ is commonly rounded up to $F_0 = 3.0$.

Environmental factors

It is well known that the nature and composition of the suspending medium will influence the rate at which organisms are killed by heat; fats, sugars and organic material in general are known to have a protective effect. With

73

pH, organisms are most resistant to thermal destruction at neutral reactions, their resistance decreasing as the suspending medium becomes acid or alkaline. It is partly from a consideration of the influence of pH on the heat resistance of bacterial endospores that the canning industry has adopted the following classification of foods:

1   low-acid foods (pH 5.0 or more)—meat products and milk;
2   medium-acid foods (pH 5.0–4.5)—soups and sauces;
3   acid foods (pH 4.5–3.7)—tomatoes, pears, etc.;
4   high acid foods (pH 3.7 and below)—citrus fruits, rhubarb, etc.

These categories also permit a prediction of the possible outcome of a process which failed to kill the spores of *Clostridium botulinum*. Until recently the available evidence indicated that this organism was unable to grow at pH 4.6 or less; the foods of categories 3 and 4 were considered to be 'safer' than those of 1 and 2 and the processing methods were adapted to kill those organisms which will cause spoilage. Recent observations (Table 4.1) indicate that a more cautious view ought to be taken about the

Table 4.1.   pH and growth and toxin production by *Clostridium botulinum* A and B (from Smelt *et al.* (1982) *J. Appl. Bact.* **52**, 75–82).

Growth and toxin production occurred in pasteurized soy protein contaminated with aerobic spore formers at pH 4.2–4.4

Toxin production was observed in a suspension of skim milk powder adjusted to pH 4.4–4.5 with HCl

Toxin production in soy protein suspensions adjusted to pH 4.4 was noted after:
    14 weeks when acetic acid was present
    12 weeks when lactic acid was present
    4 weeks when either citric or hydrochloric acids were present

supposed safety of foods of pH 4.5–3.7. As no spore-forming bacteria will grow at pH 3.7 or less, the process need only free the product of yeasts and other aciduric microorganisms. With foods having pH values of 3.7–4.5, the spores of the thermophilic *Bacillus coagulans* and mesophilic *B. macerans*, *B. polymyxa*, *Clostridium butyricum* and *Cl. pasteurianum* have to be inactivated by the heat process. In foods with a pH of 4.5 or more, the process needs to ensure not only the destruction of spores of spoilage organisms but of *Cl. botulinum* as well. Of the spoilage organisms, the thermophilic *Cl. thermosaccharolyticum*, *B. stearothermophilus* and *Desulfotomaculum nigrificans*, and mesophilic *B. licheniformis*, *B. subtilis* and *Cl. sporogenes* are of general importance (Fig. 4.3). Providing the temperature does not rise above 38 °C during the storage of a canned food, spores of *B. stearothermophilus* will remain quiescent and those of *Cl. sporogenes* will need to be destroyed if spoilage is to be avoided. As this organism is more heat resistant than *Cl. botulinum*, processing to avoid spoilage will ensure freedom from *Cl. botulinum* also. Indeed, the organism Putrefactive Anaerobe 3679, closely related to *Cl. sporogenes*, is commonly used to check the efficacy of

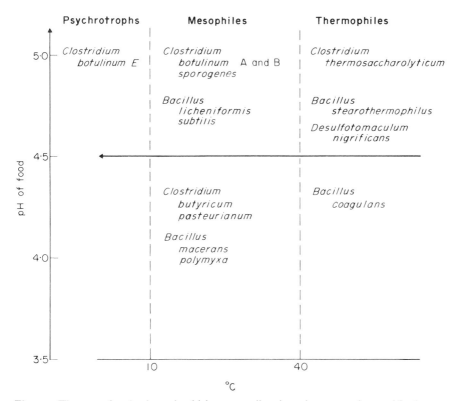

Fig. 4.3. The spore-forming bacteria which cause spoilage in underprocessed canned foods stored in the temperature ranges shown in the Figure. It will be noted that the pH of the food also plays an elective role.

a thermal process in foods at this pH. In any case the heat treatment of a food with a pH of 4.5 or more must ensure freedom of the product from contamination with *Cl. botulinum* and thus a 12 *D* process must be regarded as essential, unless effective preservatives are present.

In addition to information about the pH of a food, its heating characteristics must also be considered—is it heated by conduction alone, by convection or, due to changes occurring during heating, convection followed by conduction? This information is required if thermocouples are to be sited at the appropriate location—that part of the canned product which is the last to reach the temperature used in a process—when a retorting process is examined. With foods heated by conduction, the thermocouple will be located at the geometric centre of the container; with convection heating, it will be positioned in the central axis but towards the bottom of the can—the actual position being determined by the size of the can (Fig. 4.4).

In practice, of course, the choice of processing temperatures and times must strike a balance between the competing objectives of: (1) ensuring that the food is appertized; (2) ensuring that enzymes capable of causing organoleptic change are inactivated; and (3) conserving the flavour, colour,

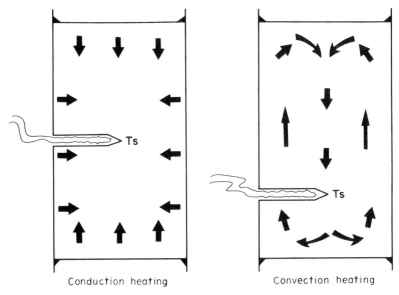

| Conduction heating | Convection heating |

Fig. 4.4. The siting of thermal sensors (Ts) in cans used to study the processing of foods heated by conduction or convection.

texture and nutritive value of the food. As changes occurring in nutrients, enzymes and endospores subjected to heat generally obey first-order reaction kinetics, D and z values can be applied generally. Thus if a processed food will be stabilized only when enzymes are inactivated, a process designed to inactivate the most resistant enzyme will ensure stability. In practice, peroxidases or catalases are selected for this purpose because both have high z values (Fig. 4.5). The information given in this Figure shows also that the z values for destruction of nutrients and quality traits are larger than those for microorganisms and enzymes. In other words a given increase in temperature causes a larger increase in the rate of destruction of micro-organisms than it does in the rate of destruction of nutrients or quality traits. This principle is exploited in HTST (temperatures of 136 °C for a few seconds or more).

Choice of processing conditions

The preceding discussion covered the general factors affecting the heat inactivation of bacterial endospores; for convenience it gave emphasis to inactivation rates and their relationship at specified temperatures. In the routine appertization of foods, a temperature range rather than a temperature will obtain and thus attention has to be given to the time taken to attain a process temperature and the time required to remove heat from a product. In practice, fluid foods permit the most rapid attainment of a processing temperature because their physical properties permit the application of a thin, moving film to efficient heat exchangers. It is for this reason that HTST is most easily applied to liquids: milk, soups, etc. When foods are packed

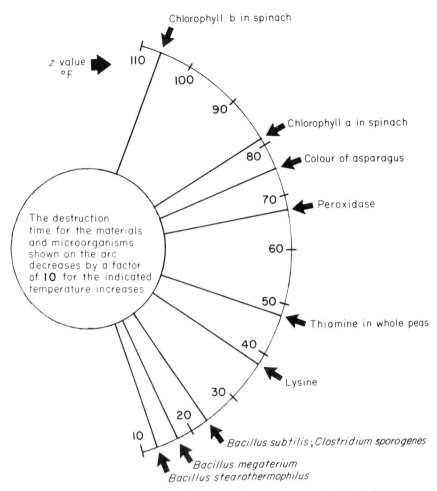

Chlorophyll b in spinach

z value
°F

110

100

90

Chlorophyll a in spinach

80

Colour of asparagus

70

Peroxidase

60

The destruction
time for the materials
and microorganisms
shown on the arc
decreases by a factor
of 10 for the indicated
temperature increases

50

Thiamine in whole peas

40

Lysine

30

20

10

Bacillus subtilis; *Clostridium sporogenes*

Bacillus megaterium
Bacillus stearothermophilus

Fig. 4.5. A comparison of the z values of microorganisms, enzymes and food ingredients.

in metal cannisters (cans), glass jars or flexible pouches, there will be a point in the container which will be the last to attain a specified temperature (Fig. 4.4). It will be the point also from which heat loss will be most sluggish. A typical heating–cooling curve is shown in Fig. 4.6. In practice the rates of temperature change at this 'cold point' will dictate the choice of time–temperature combinations for appertization. There are three broad approaches to the establishment of an acceptable thermal process: (1) biological, (2) the general method, and (3) the formula method. Number one may well be used to check 2 and 3.

(1) BIOLOGICAL

In one approach, a number (e.g. 100) of cans of food are inoculated with a known number (10 000 per can) of spores of known heat resistance, processed and incubated so that the incidence of spoiled cans can be established. As the initial spore inoculum is many times larger than that which would be expected to occur in commercial practice, the 100 test cans can be taken to be the equivalent of 10 000 commercially processed cans and the probability of spore survival under commercial conditions predicted.

In another approach, each of 10 cans is inoculated with $3-5 \times 10^7$ spores. Each can is exposed to a temperature for a known time and the number of surviving spores enumerated. The information can be used to calculate a D value in the food to be processed and thus the calculation of an Integrated Sterilizing Value (IS) for commercial use, viz.:

$$I \cdot S = D(\log I - \log S)$$

where I = the initial level and S the surviving level of contamination of a product with viable endospores.

The utility of this method will depend upon the accuracy of predictions which have to be made about (a) the levels of contamination of the product to be processed, and (b) the heat resistance of the contaminants.

(2) THE GENERAL METHOD

This method is based on the concept of the lethal rate which is:

$$\frac{1}{\text{TDT}_t} = 10^{(t - 250)/z}$$

where TDT = the thermal death time, t = the specified temperature, and z = the slope index of the thermal death time curve (°F: Fig. 4.2). In practice, a temperature-sensing device (e.g. a thermocouple) is located at the 'cold point' of the canned food, heat is applied to the container and the temperature at the 'cold point' recorded at frequent intervals until it is within $1-2$ °F of the temperature being applied. At this time, water is used to cool the container and temperatures at the 'cold point' recorded with time. Lethal rates, calculated from the formula given above and processing temperatures, are plotted against time to give a curve of the type shown in Fig. 4.6. The area under the curve corresponding to a value of 1 is referred to as the 'Unit sterilization area'. It represents a process equivalent to 1 min at 121 °C. As most thermal processes are based on the destruction of endospores having values greater than 1, total process must give an area under the curve of F (sterilizing value of heat process in minutes) × the Unit sterilization area. The area beneath the curve in Fig. 4.6 can be determined by a planimeter and an area of 10 × the Unit sterilization area is often adopted for a thermal process.

(3) THE FORMULA METHOD

The following equation provides the basis for this method:

$$B = f_h \log (j_h I_h/g).$$

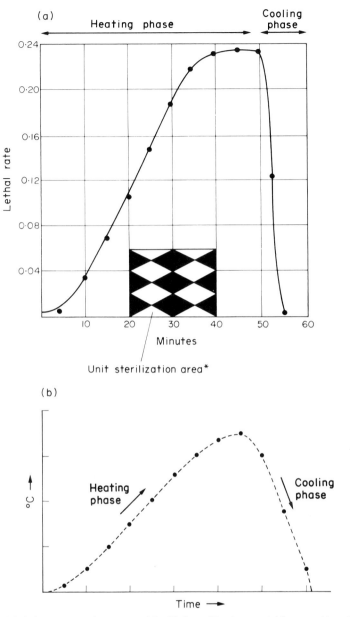

Fig. 4.6. (a) A lethal rate versus time curve with a Unit sterilization area* (chequered inset) of $F_0 = 1$ min (lethal rate of 0.05 × 20 min). (b) The corresponding heat penetration—cooling curve.

The process time (B) can be calculated when the following are known: (1) the slope of the heating line (f_h); (2) the retort temperature minus the initial product temperature $(RT - T_{ih})$; (3) the factor identifying the thermal lag $(j_h = (RT - T_{pih}'RT = T_{ih}))$, and (4) the retort temperature minus the 'cold point' temperature at the end of the heating process (g). (1), (2) and (3) can be determined from a heat penetration curve. g is not easily obtained because

79

it is a function of several factors: (1) the slope of the heating curve (f_h), (2) time of the process (expressed as the TDT) of the microorganism at the retort temperature, (3) z value, and (4) the driving force for cooling (retort temperature minus the temperature of the cooling water).

RADIATION

Ionizing radiations such as gamma rays (wavelengths, <2 Å) emitted from the excited nucleus of ^{60}Co or electrons emitted from a hot cathode and accelerated to a very high velocity have been studied extensively with the objective of producing foods that are free of spoilage microorganisms (Radappertization) or pathogens (Radicidation) or contain a greatly diminished content of spoilage organisms (Radurization). Other potential uses of ionizing radiations are given in Table 4.2. In general terms, the

Table 4.2. Uses of ionizing irradiation in the food industry.

Sterilization of foods in hermetically sealed packs

Reduction of the size of the spoilage flora on perishable foods

Elimination of pathogens in foods

Control of infestations in stored cereals

Prevention of sprouting of potatoes, carrots, etc.

Retardation of development of picked mushrooms (i.e. growth of the stem and opening of the cap delayed)

radiation of foods can be considered to be a means of achieving 'cold sterilization'—a food is freed of microorganisms without the need for high temperatures. Indeed irradiation is most commonly applied to chilled or frozen foods so that off flavours do not occur. Although ionizing radiations have an enormous potential to initiate a huge number of chemical changes in gases, liquids and solids, the splitting of water has a pivotal role in treated foods, because of its ubiquity as an ingredient. The following are formed during the exposure of water to ionizing radiation:

hydrated electron e^-_{aq}
free radicals $OH\cdot$ and $H\cdot$
excited water $(H_2O)\star$
ionized water molecules $(H_2O)^+$

Interactions between these products, or these products and other components of a food containing oxygen result in the formation of highly reactive entities (Fig. 2.6), such as hydrogen peroxide, which are toxic to microorganisms.

In practice, the nature of the food (i.e. a suspending medium's influence on an organism's sensitivity) and the properties of an organism exposed to radiation have a direct bearing on the extent of treatment required to achieve one or other of the objectives noted above. The actual treatment is measured as rads (1 rad = the absorption of 10 μJ g^{-1} of matter), kilorad or megarad. In practice, radiation effects are related to dose:

$$n = n_0 \, e^{-D/Do}$$

where n = the number of live organisms following irradiation; n_0 = initial number of organisms; D = dose radiation received (rads), and Do = a constant depending on type of organism and environmental factors.

With organisms in general, the following descending rank of sensitivity to irradiation obtains:

10^{3-4}–10^7 rads*—microorganisms killed;
10^3 –10^5 rads —insects killed;
10^3 –10^4 rads —sprouting of potatoes, carrots, etc., inhibited;
10^2 –10^3 rads —lethal dose for humans.

* Gray is the modern unit of measurement; 1 Mrad = 10 kGy.

With microorganisms, the following list ranks major groups in descending order of irradiation resistance (approximate sterilizing doses in parenthesis):
1   bacterial endospores (c. 3.0×10^6 rads);
2   yeasts (c. 5.0×10^4);
3   fungi (c. 5.0×10^4);
4   Gram-negative bacteria (c. 1–10×10^4).

There are microorganisms with exceptional resistance to irradiation, for example *Deinococcus* (*Micrococcus*) *radiodurans*, which through its highly efficient mechanisms for repair of DNA damaged by ionizing irradiation has a resistance up to 55 times greater than that of *Escherichia coli*.

Although food microbiologists have recognized the potential of ionizing radiation in food preservation, the system has been adopted very slowly due to: unacceptable changes in the flavour of some treated foods; fears that radioactivity might be induced in a food; and, perhaps of greatest importance, the concept that irradiation 'adds' something to a food such that a treated food has to be considered in the same light as one in which a preservative has been included. However, recent reappraisal by the World Health Organization and the International Atomic Energy Authority has resulted in the recommendation that radiation now be regarded as a process rather than an 'additive', and doses up to 1 Mrad be approved without the need for extensive toxicological clearance.

ASEPSIS

There are at least three situations which can be considered under the heading, asepsis—the prevention of contamination of a sterile or appertized food with organisms which can cause spoilage:
(a)   the handling of produce such as eggs, fruit, ware vegetables, etc., so that the efficiency of the *integument* as a barrier to microbial invasion is not impaired;

(b)   the construction and handling of a container so that foods which have been appertized in it are not reinfected;

(c)   the aseptic packaging of appertized foods.

Integuments

With apples, citrus fruits, etc., an undamaged skin or rind provides a barrier between the sterile tissues and the microorganisms in the environment and thus the successful storage of such products is dependent upon their biological integrity being maintained. The following will contribute:

(i)   the control of microbial pathogens and insects which damage the integument during the growth of the crop;

(ii)   methods of harvesting which minimize mechanical damage of the integument;

(iii)   grading so that produce having damaged integuments is not stored;

(iv)   packing so that the integument is not damaged during the movement of produce—e.g. the cupped trays which are used for storing apples;

(v)   application of antimicrobial agents to the integument or the packaging materials;

(vi)   storage at low temperatures with CO enrichment of the atmosphere.

Eggs

The washing of the shell of the hen's egg provides an interesting example of a cleaning agent, water, becoming a vector not only for microorganisms but also for a microbial 'nutrient' if the washing machine is of poor design or operated inefficiently. The shell is perforated with upwards of 1.0×10^4 pores having the shape of a golf tee (Fig. 4.7). In the majority of eggs laid by the domestic hen, the outer, companulate orifice is covered and crudely plugged with cuticle, a glycoprotein which is laid down in small spheres. In such eggs, however, there are upwards of 20 pores which are not capped or plugged and the incidence of such pores can be increased if the shell is scratched, as often happens when it rolls down the inclined, wire-netting floor of battery cages. A few hens in a flock lay eggs which are lacking cuticle, either completely or from one or other pole. If a warm egg is placed in cold water, the yolk and white contract more than does the shell and the pressure differential causes water and bacteria to be drawn through the pores, the bacteria becoming lodged in the shell membranes. The number of pores flooded with water is related to the 'quality' of the cuticle; shells clothed in a thick cuticle have only a few pores flooded.

Many hundred pores are flooded in those shells which were without cuticle at oviposition. If the water contains Fe^{3+} in concentrations of $2-5$ $p/10^6$, the bacteria in the shell membranes (Figs 4.8 and 4.9) will be provided with the 'nutrient' which neutralizes the principal component, ovotransferrin, of the antimicrobial defence of the albumen. The bacteria will grow rapidly and heavy contamination of the albumen results in the

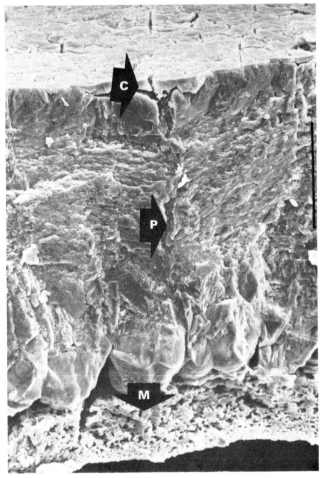

Fig. 4.7. A radial section of a pore (P) in the eggshell of the domestic hen. The outer orifice of the pore is crudely plugged with cuticle (C), a layer of spheres of glycoprotein. The inner, undulating surface of the shell is covered with two shell membranes (M; details shown in Fig. 4.8.). The bar marker = 100 μm. (The photograph was kindly given by Nick Sparks.)

addling of the egg. From this discussion, it is obvious that the following factors should be considered if the washing of eggs is to be undertaken on a commercial scale.

1 Select breeds of laying hens which produce eggs whose shells are covered with a thick cuticle.

2 Operate the washing machine so that a temperature differential is maintained between the egg (cold) and water (hot) at all stages of washing and rinsing.

3 Ensure that ferrous metals of the washing machine do not contaminate the wash water with Fe^{3+}.

4 Analyse the water for its content of Fe^{3+}.

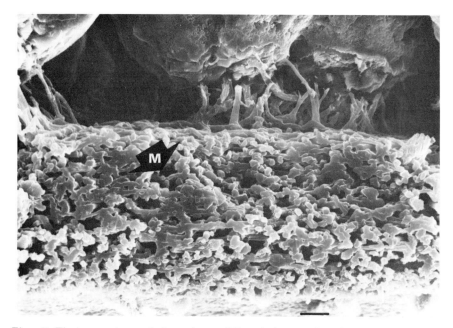

Fig. 4.8. The inner and outer shell membranes (M) on the inner surface of the eggshells of domestic hens are composed of interconnected fibres (for details see Fig. 4.9). The bar marker = 10 μm. (The photograph was kindly given by Nick Sparks.)

Cans

The successful preservation of a food by 'canning' is dependent not only on the efficacy of the thermal process but also on the design and handling of the container (Fig. 4.10) so that the processed food is protected from reinfection.

Failure to achieve the latter is referred to as post-process reinfection or microbiological leaker spoilage. The latter term has the connotation that reinfection is associated with a 'hole' in the container across which micro-organisms are translocated. As the traditional can is formed of three plates (the side and two end pieces) fixed together with joints which are soldered or filled with a sealing agent, it can be assumed that holes will normally be associated with such seams or joints (Fig. 4.11). In a detailed analysis of leaker spoilage, Put and her collaborators (1972, 1980) identified eight factors which can contribute to reinfection of canned foods:

(a) *faults* in the design or manufacture of the seams in the can (see Figs 4.12 and 4.13);

(b) the *geometry* of a hole extending across a seam;

(c) the *level* of contamination of water with which a can made contact (Fig. 4.14);

(d) the *duration* of contact of a seam and a contaminated liquid (Fig. 4.15);

(e) the *viscosity* of the contaminated liquid in contact with a seam; viscous liquids offer the greatest impediment to the translocation of micro-organisms;

84

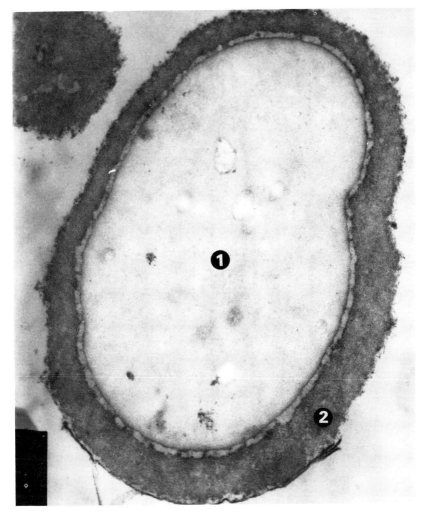

Fig. 4.9. The fibres of the shell membranes of the eggshells of domestic hens are composed of (1) a core of fibrous protein rich in desmosine and isodesmosine but resistant to elastase, and (2) a mantle of glycoproteins. The acidic groups on the mantle cause iron to be absorbed into the membranes thereby providing bacteria trapped in the membranes with a source of this element.

(f) the *level* and *incidence* of insults—collisions with other cans or with equipment—to which a wet seam is exposed (Fig. 4.16);

(g) a *negative* pressure in the can, and

(h) sudden *fluctuations* in the pressure outside the can.

As with many examples discussed in this section of the book, the policy for the day-to-day control of leaker spoilage cannot be applied to one contributory factor alone, it must take account of the entire operation. Thus from a study of the contributors to leaker spoilage listed above, it is evident that control measures have to be considered at the time that the process is being planned—to minimize collisions between cans or between cans and

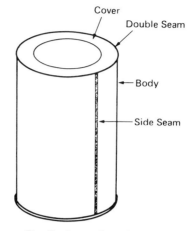

Fig. 4.10. Details of a metal cannister ('can').

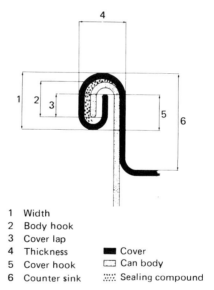

1 Width
2 Body hook
3 Cover lap
4 Thickness   ■ Cover
5 Cover hook   ▭ Can body
6 Counter sink   ░ Sealing compound

Fig. 4.11. Details of the double seam (see Fig. 4.10) that is formed when a can is closed with a metal lid.

equipment so that seams are not deformed. The cans must conform to a standard and the lidding-machine set so that the seams are effectively closed. When these measures are taken then the aseptic handling of the cans has to be considered:

(i) chlorinate $(4-5 \ p/10^6$ chlorine) the cooling water and allow contact for at least 20 min—according to Put *et al.* (1972), the water drained from the seam of a cooled can should contain $1-2 \ p/10^6$ of free chlorine;

Faults in the Double Seam

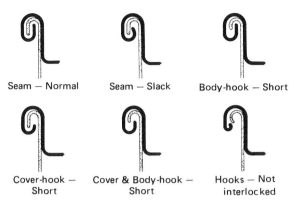

Seam — Normal   Seam — Slack   Body-hook — Short

Cover-hook —   Cover & Body-hook —   Hooks — Not
Short   Short   interlocked

Fig. 4.12. Bad adjustment of the lidding device can cause malformed double seams when a can is closed with a metal lid.

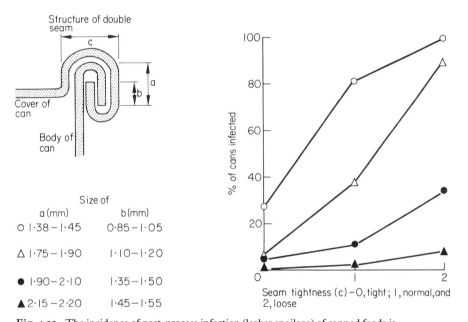

Structure of double seam

Cover of can

Body of can

Size of

a (mm)	b (mm)
O 1·38 – 1·45	0·85 – 1·05
△ 1·75 – 1·90	1·10 – 1·20
● 1·90 – 2·10	1·35 – 1·50
▲ 2·15 – 2·20	1·45 – 1·55

% of cans infected

Seam tightness (c) – 0, tight; 1, normal, and 2, loose

Fig. 4.13. The incidence of post-process infection (leaker spoilage) of canned foods is determined in part by the dimensions of the double seam (see Fig. 4.11) and the extent to which it is pressed when the can is closed. (Based on details from Put *et al.* (1972).)

(ii)   clean and disinfect at frequent intervals equipment with which the double seam makes contact; monitor routinely the effectiveness of the cleaning operation;

(iii)   ensure that cans are dried thoroughly before storage;

(iv)   ensure that the workers are aware of those practices which contribute to leaker spoilage.

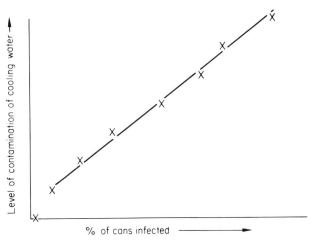

Fig. 4.14. The incidence of leaker spoilage of canned foods is influenced by the level of contamination of the water used to cool the can post-retorting. (Based on details from Put *et al.* (1972).)

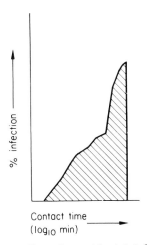

Fig. 4.15. The incidence of leaker spoilage of canned foods is influenced by the length of time that a can is exposed to microorganisms suspended in water. (Based on details from Put *et al.* (1972).)

As canned foods are intended to have a long storage life, the storage conditions must be such that the integrity of the cannister is maintained. Storage at high temperature can lead to corrosion of the inner surface of the can. If the humidity of the store is too high, the can may rust. Rusting will be accentuated if the label is made of paper containing too much Cl^- or if it is attached with glue containing oxidizing agents.

Post-processing contamination could well be a problem with foods appertized in flexible pouches which do not have the puncture resistance of metal or glass containers. A recent survey showed that 'rough' as opposed to careful handling during the packing of a retort with flexible pouches

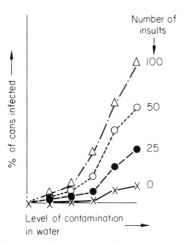

Fig. 4.16. The incidence of leaker spoilage of canned foods is influenced by the number of times a filled can collides with another can or parts of the conveyor belt, etc., (insults), as well as the level of contamination of the cooling water. (Based on details from Put *et al.* (1972).)

increased the percentage (0.06–0.27%) of punctured containers. A 90% contamination rate was found when deliberately punctured pouches were removed manually from a retort, cooled in tap water and stored wet; a 10% rate obtained when such pouches were dried before storage and a 1% rate was recorded when punctured pouches were cooled (with chlorinated water) and dried in the retort. Such evidence points yet again to the need to use water of good quality for cooling food in containers and to provide adequate training for and supervision of the persons engaged in the handling of appertized foods. Moreover, it must be recognized that through being flexible such pouches may cause a pumping action such that microorganisms are sucked through holes in the walls of the pouch.

Aseptic packaging

There are two approaches to the aseptic packaging of a product that has been appertized by HTST. The processed food may be filled into unsterilized containers at high temperature (124 °C), the filling being done in a pressurized chamber (18 p.s.i.). After the container has been closed, it is retained in the chamber until contaminants have been inactivated. Alternatively, sterilized containers are fed through locks into a chamber containing the apparatus for dispensing the processed food and sealing the container (Fig. 4.17). The inside of the chamber has to be sterilized before the start of production; this is commonly achieved with superheated steam at atmospheric pressure. The germ-free status of the chamber needs to be maintained during production, a problem of course as locks are required for the admission of sterile containers and the discharge of filled ones. The use of superheated steam and air at a positive pressure within the chamber will aid the maintenance of sterility. In order that oxidative changes do not

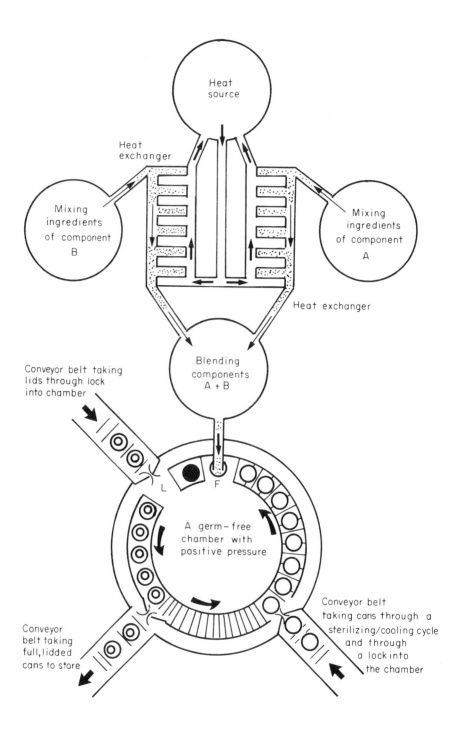

Fig. 4.17. Hypothetical layout of an aseptic canning process in which two major components (A and B) are sterilized separately, mixed and filled (F) into sterile cans resting on a race-track conveyor belt within a germ-free, pressurized chamber. The chamber contains a lidding device (L) for closing the cans and ports for the admission of sterile cans and lids and the exit of finished products.

impair the organoleptic qualities of a canned product, there may be a need to remove oxygen from the head space of the container before the lid is sealed on. Culinary quality steam or nitrogen can be used for this purpose. The material properties of the containers will dictate the methods used for their sterilization. With cans, for example, sterilization is achieved by passing them along tunnels to which superheated steam (500–600 °F) is fed; the sterilized cans may need to be cooled with sterile water in order to prevent blistering of the enamel on the surface of the metal. With heat-labile materials such as plastic or cardboard, H_2O_2, UV light or chlorine are used for sterilization.

5 Deliberate Infection

The opposing strategies of our forefathers in their quest for methods of preserving foods were noted in Chapter 2; they attempted to inhibit or destroy organisms or they encouraged the growth of those that produced a food having acceptable qualities and, in most cases, an extended storage life. The long history of the latter approach—there is evidence that the Chinese have practiced fermentation for the past 3000 years—means there is a notable diversity in the primary objectives of fermentation processes and an equally diverse range of products that have been adopted for man's pleasure or nutrition. This is exemplified by the information given in Table 5.1. It highlights also the influence of culture; in general, bacteria

Table 5.1. Fermented foods.

	Substrate or product
Fermentations due to bacteria	
Acetobacter spp.	Oxidation of ethanol to acetic acid in vinegar production
Zymomonas mobilis subsp. *mobilis*	Plant juices used in alcohol production
Streptococcus, Lactobacillus and *Leuconostoc*	Production of lactic acid in cheese and sauerkraut production
Pediococcus halophilus	Acidification of soy sauce
Bifidobacterium spp.	Fermentation of milk
'*Lactobacillus sanfrancisco*'	Leavening of sour dough bread
Lactic acid bacteria and *Staphylococcus carnosus*	Fermented meat products, e.g. salami
Fermentations due to fungi and yeasts	
Saccharomyces cerevisiae	Production of alcohol / Leavening of bread
Saccharomyces rouxii	Fermentation of soy sauce
Aspergillus oryzae	Breakdown of starch in rice grains
Rhizopus oligosporus	Removal of the bean flavour from soy bean
Monascus purpureus	Pigmentation of rice grains

dominate the fermentations in the West whereas moulds have been favoured in the Orient.

Many reasons can be identified tentatively for the adoption of techniques to ferment foods. The primary objective may be to preserve a seasonal crop for use throughout the year; the fermentation of cabbage (sauerkraut), small cucumbers and olives can be sited as examples. Likewise the production of fermented milk products is basically a method of ensuring the storage of the major nutrients of a very perishable material. It will be noted below that a very wide range of cheeses are produced, often by encouraging the growth of organisms that bring about desirable changes of the material produced during the initial fermentation. Thus preservation is linked with practices that cause changes in the organoleptic properties of products. In other cases,

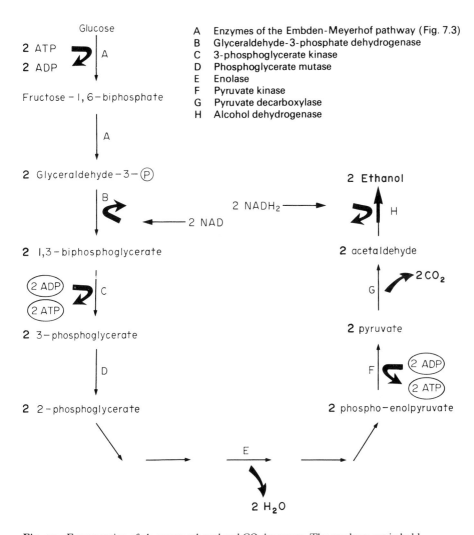

Fig. 5.1. Fermentation of glucose to ethanol and CO_2 by yeasts. The products are in bold.

fermentation results in an acceptable commodity being produced from raw materials having unacceptable properties. Thus fermentation removes protease inhibitors, flatulents and the 'bean flavour' of soybeans during the production of tempeh. Bread is leavened through the production of CO_2 from the fermentation of maltose by *Saccharomyces cerevisiae*. '*Lactobacillus sanfrancisco*', *S. exiguus* and *S. inusitatus* are responsible for fermenting carbohydrates in sour dough bread; the heterofermentative lactobacillus lowers the pH of the dough to 3.8–4.0 through the production of lactic acid, acetic acid and ethanol from maltose. Indeed aesthetic rather than nutritional considerations can be identified as important objectives in certain fermentations, as in the fermentation of rice with *Monascus purpureus*. The polished rice is stained purplish-red as a result of the mould producing red (monascorubin) and yellow (monascoflavin) pigments. The stained grains are dried and ground to a flour (ang-kak) which is used to colour and flavour foods prepared from fish or cheese. The pigmented grains are used also to produce red rice wine. The adoption of a catholic definition of fermentation would include beers and wines, products produced from the fermentation of plant juices or grain extracts by *S. cerevisiae* (Fig. 5.1) or *Zymomonas mobilis* subsp. *mobilis* (Fig. 5.2). A spoilage product, acetic acid, of a failed alcoholic fermentation was used traditionally for food preservation. There are also prospects of fermenting carbohydrate-rich plant materials so that protein enrichment occurs and a better diet can be offered to people on a subsistence standard of living.

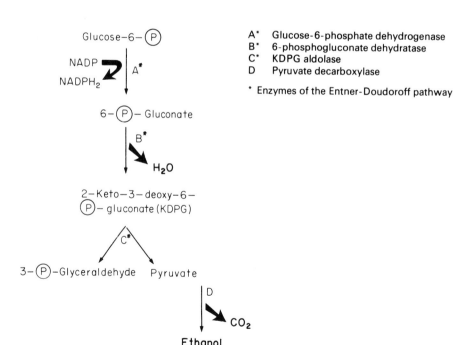

Fig. 5.2. Ethanol production by *Zymomonas mobilis* subsp. *mobilis*. The products are in bold.

THE PRODUCTION OF ALCOHOL

Ethanol (alcohol) is the essential component of alcoholic beverages such as cider, perry and wines or potable spirits distilled from them. In most cases yeasts, *Saccharomyces cerevisiae* or *S. uvarum*, ferment sugar-rich solutions of nutrients obtained from fruit (apple = cider, pears = perry, grapes = wine), extracts of seeds (e.g. barley) or other plant material (e.g. potatoes). With the last two examples, fermentable substrates are formed from the starch reserves by endogenous or added amylases. Another yeast, *S. beticus*, is associated with the production of fortified wines such as Spanish sherry.

With other alcoholic beverages, bacteria as well as yeasts contribute to the fermentation. Thus some locally brewed South African beers are produced by a fermentation to which lactic acid bacteria contribute. The fermentation is, therefore, analogous to that occurring during the production of sour dough bread (see p. 94). Tibi is an acidic, mildy alcoholic drink prepared by the fermentation of sucrose solutions containing figs, dates, raisins, lemons or ginger as a source of flavour and the nutrients required for the growth of a consortium of microorganisms, *Lactobacillus brevis*, *Streptococcus lactis* and *Saccharomyces cerevisiae*. The microorganisms form grains with a dextran matrix consisting mainly of a backbone of α-D(1-6)-linked glucopyranosyl residues with (1-3)-linked side-chains. Thus Tibi grains, which occur as hard granules on the leaves of opuntia plants (prickly pear cacti), are similar to the Kefir grains discussed on p. 7.

In some parts of the world alcoholic beverages are produced through the fermentation of plant juices by a bacterium, *Zymomonas mobilis* subsp. *mobilis* (Fig. 5.2). In Mexico, for example, the fermentation of the sugary sap (aguamiel) of the *Agave* (the Amaryllis family of plants) by this organism is a part of the process leading to the production of mescal, pulque or tequila. It is considered to play an important role also in the production of palm wines throughout the tropics.

The following description of beer production identifies the various stages that result in a flavoured alcoholic beverage being produced from a seed, barley, in which starch is the main reserve of carbohydrate.

(a) Malting and mashing

The quiescent barley seed has to be germinated (malted) so that its content of endogenous enzymes is increased to such an extent that polymers extracted from the seed are quickly digested during the mashing process. Of the cereal seeds available commercially, barley is preferred because the husks protect the germinating seed from mould infection and act as a 'filter aid' during the aqueous extraction process. Moreover, the gelatinization temperature of barley starch is lower than that which inactivates the α-amylase contained in the seed. Thus gelatinization and solubilization of the starch can be achieved in one operation. The following are the main biochemical objectives of the malting process: (i) to encourage the synthesis of endo-β-glucanase, α-amylase and peptidases, and (ii) to digest part of the walls of the cells of the endosperm so that yeasts are provided not only with fermentable carbohydrates but peptides and amino acids for growth.

The objectives noted above are achieved by a controlled germination (malting) of the barley—germination is stopped at the desired stage and the barley dried (kilned). The germinated (malted), dried barley seed is ground and extracted with water of up to 67 °C (the mashing process). Rapid degradation of solubilized starch ensues. The extract—'sweet wort'— is boiled with hops (these impart the bitter flavour to beers) and, after being allowed to settle, the supernatant (hopped wort) is cooled.

(b) Pitching

The hopped wort is inoculated ('pitched') with a selected strain of *Saccharomyces cerevisiae* or *uvarum*. Growth of the yeasts and fermentation (Fig. 5.1) of sugars, mainly maltose and glucose, in the wort causes a diminution of gravity—attenuation is the term used to describe this change in the industry. Once attenuation has reached a limit, the yeasts are separated, the 'green-beer' is allowed to mature before filtration, pasteurization and dispatch in bottles, cans, barrels or kegs.

THE PRODUCTION OF VINEGAR

A wide variety of carbohydrate-rich plant materials are used for vinegar production, malted barley in the UK, apples in the USA, grapes in France, etc. Such materials are fermented by yeasts in order to produce ethanol, the substrate used by the acetic acid bacteria of which the genus *Acetobacter* is of primary importance in vinegar production. Organisms of this genus oxidize ethanol to acetic acid and, when ethanol is no longer available, acetic acid to CO_2 and water. Both oxidations are exothermic.

As oxygen is essential for the conversion of ethanol to acetic acid, the acetobacters in contact with ethanol have to be provided with an adequate supply of air. The means whereby this is achieved are discussed below.

The Orleans process

Wine undergoing acetification (i.e. ethanol is being oxidized by acetic acid bacteria) is used to partly fill a cask. The microorganisms form a pellicle which can be mechanically strong if the cellulose-producing *Acetobacter xylinum* is present—the cellulose fibres become matted together. When the ethanol has been oxidized to acetic acid, about half of the acetified wine is drawn off and replaced with fresh wine.

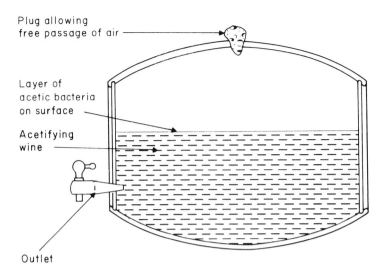

Plug allowing free passage of air

Layer of acetic bacteria on surface

Acetifying wine

Outlet

The trickling filter process

Fermented beer wort ('gyle') is allowed to trickle through a bed of wood shavings contained in a wooden tower with a perforated shelf several feet

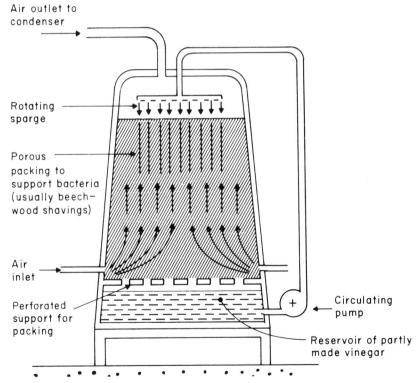

Air outlet to condenser

Rotating sparge

Porous packing to support bacteria (usually beech–wood shavings)

Air inlet

Perforated support for packing

Circulating pump

Reservoir of partly made vinegar

These diagrams and that on p. 98 redrawn from J.G. Carr (1982) The production of foods and beverages from plant materials by microorganisms. In *Bacteria and Plants* (Eds. Rhodes-Robert M.E. & Skinner F.A.). Society for Applied Bacteriology Symposium series No. 11.)

above the foundations. Air holes with shutters are situated in the side of the tower and just above the shelf. Acetic bacteria colonize the surface of the wood shavings, oxidize the ethanol to acetic acid and produce heat so that the tower acts as a chimney thereby ensuring an updraught of air through the shavings. Indeed the acetification process can be controlled by the rate of flow of air through the shuttered ports in the side of the tower. As with the trickle filters discussed on p. 165 excessive growth of acetic acid bacteria, especially *Acetobacter xylinum*, can fill the void spaces in the shavings thereby impeding the flow of wort and air.

The submerged process

This process is based on a technique familiar to all students of microbiology. Acetic acid bacteria are suspended in a nutrient medium containing ethanol which is sparged with air. As the oxidation of ethanol is an exothermic reaction, a thermostatically controlled cooling coil is fitted to the fermenter so that temperatures injurious to the acetic acid bacteria are never achieved. A spinning disc in the head space of the vessel prevents excessive foam formation. Care must be taken to ensure that the air used to sparge the fermenter is free of antimicrobial agents; problems can be encountered, for example, when fermenters are sited in industrial areas where SO_2 is present in the atmosphere.

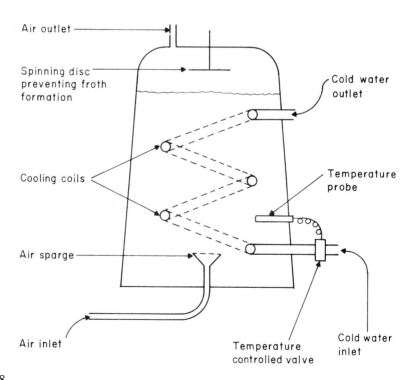

PROCESSING OF COCOA AND COFFEE BEANS

Not only is fermentation used for the traditional purposes noted above but also in the preparative stages of raw materials, such as cocoa and coffee beans, whose ultimate qualities and properties in a finished product are determined in large part by post-fermentation processing. Thus in practice, micro-organisms are used at a very early stage in the processing of a food or beverage ingredient in much the same manner as *Clostridium felsineum* or *Chaetomium* spp. are used to release fibres during the retting of flax.

When a cocoa crop is harvested from the trees (*Theobroma cacao*), 30–40 beans each having two bright purple cotyledons surrounded by a pink testa are embedded in the mucilage contained in a pod of 25 cm length and 12 cm diameter. The beans scooped from the pods are either piled into heaps and covered by plaintain leaves or put into perforated boxes so that a fermentation can occur, the microorganisms probably deriving most of their nutrients from the mucilage. When judged by an analysis of the fermenting flora, a complex fermentation is involved in the production of mucilage-free beans. At the outset species of yeasts of the genera *Kloeckera*, *Hansenula* and *Saccharomyces* produce ethanol which, in turn, leads to acetic acid production by *Acetobacter* spp. As these two fermentation products accumulate and the sugars in the mucilage become depleted, lactobacilli, for example *Lactobacillus plantarum*, come to the fore and the pH drifts to a final value of *c.* 5.2. Not only are the beans freed of mucilage, but the embryo is killed by the heat (upwards of 50 °C) of fermentation, the cotyledons change in colour due to the liberation of anthocyanins and oxidation of polyphenols, and the first vestiges of a chocolate flavour and aroma become evident. Similar changes may occur also during the fermentation of the mucilage-coated beans taken from the coffee berry. In this case, '*Erwinia dissolvens*' may be involved in the depolymerization of the mucilage.

FERMENTATION—A BROAD DEFINITION

Those who have studied microbial physiology will have deduced from the above that fermentation has a broader definition in food microbiology than it does in physiology: an ATP-generating process in which organic compounds serve as electron donors and acceptors. Indeed when the metabolic attributes of the range of organisms associated with food fermentations are considered, it is apparent that in addition to organisms which bring about fermentations *sensu strictu*, there are many others that act mainly through secreting amylases, pectinases or lipases. With the former the fermentation may be based on the activities of one organism (cottage cheese), a succession of organisms (sauerkraut) or a consortium of organisms (kefir). With the examples given in parenthesis, lactose or glucose and fructose will be readily available to the organisms. When the substrate needed for fermentation is in polymeric form in the raw material, depolymerization is an essential first step. This may be achieved by endogenous enzymes, as in the malting of barley, or by fostering the growth of an appropriate microorganism. With soy sauce, for example, the growth of *Aspergillus oryzae* on polished rice leads to the production of proteases and amylase which bring about the breakdown of proteins and starch when the rice is combined with boiled soy beans and crushed, roasted whole grains of wheat and incubated at 30 °C for 72 h. After the dilution of this mash with brine, a fermentation is initiated by *Pediococcus halophilus*. With the fermentation of glucose, the pH drifts from 7.0–6.5 to 4.5 at which value an ethanonolic fermentation is associated with the growth of *Saccharomyces rouxii*. The fermentation phase can last for one to three years. With temper, a mould, *Rhizopus oligosporus*, which produces large amounts of lipases and proteases but little amylase ensures that the major changes are associated with fats and proteins but not starch during a short fermentation phase (20–24 h at 30 °C).

Although Table 5.1 shows that there is a marked diversity in the application of fermentation in the food industry, the remainder of this chapter will be confined to two products only:

1 fermented plant materials (cabbage, cucumbers and olives);
2 fermented milk and cheese.

In the discussion of fermented plant materials, emphasis has been given to two features:

(a) problems that can arise from contamination of raw materials with macerating enzymes or the microorganisms that produce them;
(b) the evolution from a craft to a technology.

Indeed the latter would be apposite also for a discussion of fermented meat products where there has been a gradual move away from fermentations initiated by contaminants of raw materials to ones which are dependent upon the addition of starter cultures of lactic acid bacteria, particularly *Pediococcus cerevisiae*, and *Staphylococcus carnosus*.

The discussion of milk and cheese draws attention to the contributions of the biochemist and molecular biologist to a technology whose evolution from a craft-state was due mainly to the efforts of microbiologists.

The title to this chapter may be taken to mean that selected organisms are used to infect a product. Although this can be and is now done commercially, the traditional methods provide conditions which *select* particular microorganisms; the latter become numerically dominant and they may poison the environment both for themselves and any would-be pioneer of a succession, as in sauerkraut, or they may provide an environment in which another organism causes changes contributing to the taste, texture or appearance of a food, as with microorganisms which 'ripen' cheese. Thus the traditional methods may be likened to enrichment cultures having as essential features:

(i) an inhibitor of particular organisms—e.g. the brine in which cucumbers are stored;
(ii) a source of nutrients—extracted from plant materials by the brine;
(iii) a source of fermentable carbohydrates—lactose, glucose and fructose;
(iv) a poor buffering capacity so that accumulation of fermentation products causes a rapid drift in pH;

(v)   anaerobiosis so that the fermentation products are not oxidized by other microorganisms.

Those who have studied systematic bacteriology will recognize these parameters as those which can be expected to enrich for lactic acid bacteria, and the study of the two fermented foods discussed in this chapter is largely a study of such organisms.

(1)   Fermented plant materials

The producers of fermented cucumbers, sauerkraut or olives have as their major objective the rapid establishment of a fermentation of carbohydrates extracted from the plant material with brines. The fermentation must be such that the products do not have an offensive odour or taste; the products of fermentation must be conserved, and any change in the texture and colour of the product must be such that the discerning consumer associates them with quality.

THE FERMENTATION

The use of brine has three functions:
  (i)   to retain crispness of plant tissues;
  (ii)   to extract plant materials (amino acids, vitamins, glucose and fructose) so that microorganisms are provided with an adequate medium for growth;
  (iii)   to inhibit the growth of *Clostridium* and *Bacillus* spp. The inhibition of the growth of the former has been attributed traditionally to the a_w being too low—one may now ponder on the role, if any, that singlet oxygen arising from disturbed metabolism of plant tissues may have (Fig. 2.6).

With traditional methods, there is a succession of microorganisms from those having low tolerance of acid in a brine (the coliforms) through those having a moderate tolerance (*Leuconostoc mesenteroides, Pediococcus cerevisiae*) to those having a marked tolerance (*Lactobacillus brevis* and *L. plantarum*). The rate of progression of this succession will be governed mainly by the ambient temperature, the concentration of the brine and the availability of nutrients. When the metabolic attributes of the lactic acid bacteria are considered (Table 5.2) it is obvious that lactic acid and acetic acid and CO_2 will be the major fermentation products; in other words, products having a bland flavour at the concentrations which they achieve. The initiation of the succession may be dependent upon some treatment of the material before it is placed in the fermentation vat. Thus with olives, storage (12–14 h) in lye (0.9–1.25%) extracts the inhibitors, an aglycone and elenolic acid (see p. 28), of lactic acid bacteria. Moreover, the deliberate acidification of the brine, with lactic or acetic acid, and its enrichment with fermentable carbohydrates have been done with the object of enhancing the rate of microbial fermentation. An enhancement of the rate of fermentation has been achieved experimentally by 'heat-shocking' olives, it being surmised that this treatment, through increasing the permeability of the skin on the fruit,

Table 5.2. Metabolism of lactic acid bacteria and properties of food.

Carbohydrate metabolism	Organism	Changes in food	
Glucose → lactic acid Fructose → lactic acid Pentose → lactic and acetic acid	Homofermentative lactic acid bacteria	D	Products contribute to flavour and preservation of sauerkraut, pickled olives, cheese, fermented sausages
		SP	'Souring' of many products
Glucose → lactic acid, ethanol + CO_2 Fructose → lactic acid, mannitol, acetic acid + CO_2 Pentose → lactic acid and acetic acid	Heterofermentative lactic acid bacteria	D	Products contribute to flavour and preservation of sauerkraut, pickled olives, cheese, fermented sausages
		SP	'Souring' of many products; 'bloating' of brined cucumbers
Polymer synthesis	*Leuconostoc mesenteroides*	SP	Slime in sugar refineries, cider and wines
	Lactobacillus spp.	SP	'Ropey' beers and cider
	Pediococcus cerevisiae	SP	'Ropey' cider
Organic acids Citric → acetic, formic, diacetyl, acetoin + CO_2	Homofermentative and heterofermentative lactic acid bacteria	D	Butter flavour in cottage cheese (see Fig. 5.7)
Malic → lactic, acetic, ethanol, formic, acetoin + CO_2		SP	'Bloater' formation in brined cucumber—CO_2
		SP	Diacetyl production in beer (*Pediococcus cerevisiae*)
Tartaric → lactic, acetic, succinic + CO_2		D/SP	Malo-lactic fermentation in wines and ciders
Pyruvic → diacetyl, acetoin, acetic acid, ethanol, lactic acid + CO_2		D	Butter flavour in cheese (*Streptococcus thermophilus*)
Miscellaneous Peroxide production	*Lactobacillus viridescens*	SP	Bleaching or greening of pigments in meat products
Pigment production	*Lactobacillus* spp.	SP	Brown spot in Cheddar cheese; discoloration of sauerkraut

D, desired change; SP, spoilage.

results in a rapid accumulation of nutrients in the brine. In addition it is probable that the heat treatment inactivates the β-glucosidase associated with the formation of an aglycone and elenolic acid from oleuropein (see p. 28).

Under optimal conditions, the fermentations will be dominated by heterofermentative (Fig. 5.3) and homofermentative (Fig. 5.4) lactic acid bacteria. There are descriptions in the literature of other fermentations occurring: (i) the 'stuck fermentation' of olives—yeasts achieve a dominance and ethanol and CO_2 are the principal products of fermentation (Fig. 5.1), and (ii) 'mixed fatty acid fermentations' (formic, propionic, butyric, valeric, caproic and caprylic acids) associated with the growth of *Clostridium* and *Propionibacterium* spp.

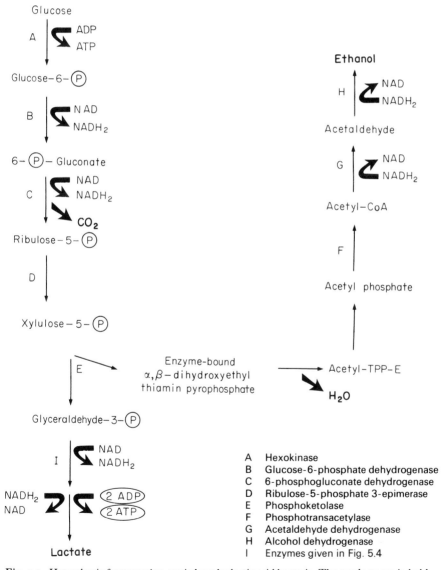

A   Hexokinase
B   Glucose-6-phosphate dehydrogenase
C   6-phosphogluconate dehydrogenase
D   Ribulose-5-phosphate 3-epimerase
E   Phosphoketolase
F   Phosphotransacetylase
G   Acetaldehyde dehydrogenase
H   Alcohol dehydrogenase
I   Enzymes given in Fig. 5.4

Fig. 5.3. Heterolactic fermentation carried out by lactic acid bacteria. The products are in bold.

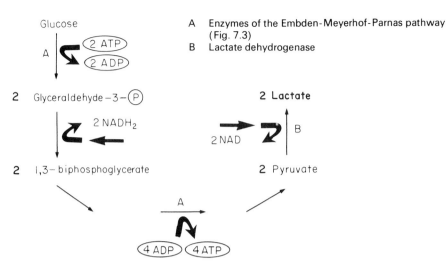

Fig. 5.4. Homolactic fermentation as carried out by lactic acid bacteria. The products are in bold.

Anaerobic respiration whereby a halotolerant *Desulfovibrio* sp. couples the oxidation of lactic acid with the reduction of SO_4^{2-} (Fig. 5.5), a contaminant of solar salt, can also lead to spoilage. The CO_2 produced during fermentation can result in spoilage as well; if the gas comes out of solution in the plant tissues, fruits are deformed, a condition referred to as 'bloaters' in cucumbers. This fault has been attributed to the growth of coliforms, yeasts and heterofermentative bacteria; it can be caused also by the homofermentative *Lactobacillus plantarum* and one might surmise that in the latter case the CO_2 is formed from organic acids (Table 5.2).

It was noted above that the successful preservation of a fermented product is dependent upon the conservation of acids formed by the micro-organisms. It is important that acid-tolerant aerobes such as yeasts do not colonize the surface of the brine and oxidize the organic acids. Colonization during the fermentation stage can be prevented by exposing the surface of the brine to UV irradiation, either from the sun or a lamp, or covering it with an oil, an inhibitor such as mustard oils or sealing the top of the fermentation vat with a plastic bag filled with water. It is important to ensure also that all the carbohydrate is used up in the fermentation stage otherwise a secondary fermentation in the packed product can lead to turbidity in the brine and a positive pressure in the container. Traditionally critical control of the salt content of the brine together with the addition of preservative amounts of vinegar and sucrose and hermetic sealing of the container assured long-term storage. Another approach is to add preservatives, such as sodium benzoate, sorbic acid or potassium bisulphite. Not only do such preservatives aid shelf-life during storage but they prevent the formation of a film of yeast on the surface of the brine in large containers of product when these have been opened in the home or canteen. Pasteurization is now widely applied because in addition to the killing of potential spoilage organisms, pentinolytic enzymes are inactivated also.

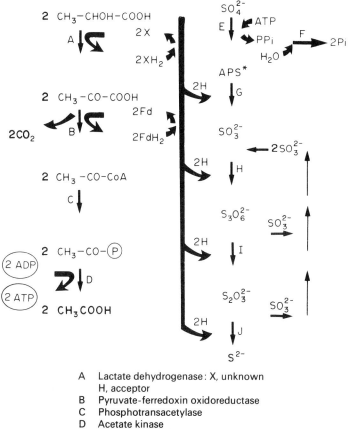

A   Lactate dehydrogenase: X, unknown
    H, acceptor
B   Pyruvate-ferredoxin oxidoreductase
C   Phosphotransacetylase
D   Acetate kinase
E   ATP sulphurylase
F   Pyrophosphatase
G   APS reductase
H   Sulphite reductase (desulfoviridin)
I   Trithionate reductase
J   Thiosulphate reductase

* Adenosine-5'-phosphosulphate

Fig. 5.5. Sulphate is used as a terminal electron acceptor by *Desulfovibrio* that cause blackening of fermented vegetables. The same pathway is used also by *Desulfotomaculum nigrificans*, a thermophile found in canned foods. The products are in bold.

THE RETENTION OF STRUCTURAL INTEGRITY

The hydrolysis of the pectin bridges between cells in tissues can lead to softening or maceration of plant materials, either during fermentation or storage. Softening has been associated with produce in which very active fermentation has resulted in the accumulation of 0.8–1.0% lactic acid. In general, however, microorganisms are the main cause of softening, moulds being of particular concern because of the acid tolerance of their poly-galacturonases. Some of the organisms which have been associated with softening or maceration are listed in Table 5.3. The table gives also the sites of growth of the pectinolytic microorganism emphasizing that methods

Table 5.3. Microorganisms which have been associated with maceration of fermented plant materials.

Organism	Site of growth	Spoilage of
Fungi		
Cytospora leucostoma	On the fruit (a sulphite-stable polygalacturonase formed)	Softening of bisulphite-brined cherries
Penicillium oxalicum		
Ascochyta cucumis		
Fusarium roseum		
Cladosporium cladosporioides	Flowers on cucumbers	Softening of cucumbers
Alternaria tenuis		
Fusarium oxysporum		
Fusarium solani		
Yeasts		
Rhodotorula glutinis subsp. *glutinus*		
Rhodotorula minuta subsp. *minuta*	Surface of brine	Softening of olives
Rhodotorula rubra		
Saccharomyces oleaginosus		
Saccharomyces kluyveri	In the brine	Softening and gas-pocket fermentation in olives
Hanensula anomala subsp. *anomala*		
Bacteria		
Bacillus subtilis	In the brine	Softening of olives
Bacillus pumilus		

of control have to be considered at *all* stages in the production of fermented plant materials. Thus with cucumbers, the growth of pectinolytic moulds on the withering flowers leads to pectinases being introduced to the fermentation vats where all the fruits will be bathed in a dilute solution of the enzymes. In this example, control can be attempted by:

(a) *selecting* varieties of cucumbers from which the withered flower falls while the crop is in the field;

(b) *grading* produce so that damaged and diseased materials are not put into the fermentation vats;

(c) *eluting* the enzymes in brine before allowing microbial fermentation to proceed in another brine;

(d) *adding* an inhibitor of pectinases to the brine in a fermentation tank.

Although extracts of various plants have been found to contain potent inhibitors of pectinases and proteinases, these have not been used commercially. The control of maceration resulting from the growth of pectinolytic microorganisms in or on brines directs attention to the methods used to promote a successful fermentation, a topic discussed on page 107.

RETENTION OF COLOUR AND APPEARANCE

It is essential to control the growth of pigment-producing microorganisms, particularly yeasts, during fermentation or the product will be stained. This happens with sauerkraut when the brine is colonized with pink yeasts. Likewise, the growth of halotolerant *Desulfovibrio* sp. in a fermentation brine can cause blackening of plant materials through the reaction of Fe^{2+} with H_2S, the reduction product formed during anaerobic respiration with SO_4^{2-}. It has been demonstrated experimentally that certain combinations of brine and pH cause *Lactobacillus plantarum*, a member of the succession associated with the fermentation of plant materials, to form a brownish-red pigment. This organism also causes 'yeast spots' on olives; it colonizes the space beneath a stomata and the colony appears as a white spot.

CONTROLLED FERMENTATIONS

Once it was realized that the successful production of sauerkraut, fermented cucumbers and olives was dependent upon the activities of microorganisms (*Leuconostoc mesenteroides, Pediococcus cerevisiae* and *Lactobacillus* spp.) which are only *minor* contaminants of the raw materials, traditional methods of production could be modified so that conditions were made more elective for these organisms. It was noted above that brines were acidified and enriched with fermentable carbohydrates. In another approach, the brines were seeded with cultures or 'starter' organisms (Table 5.4). It is

Table 5.4. Organisms used as starter cultures.

Organism	Example of foods
Streptococcus lactis subsp. *cremoris**	Cottage cheese
Streptococcus lactis subsp. *lactis*	English shire cheeses, e.g. Cheddar
Streptococcus lactis subsp. *diacetilactis*	Soured cream and buttermilk
Leuconostoc mesenteroides	Cottage cheese
Streptococcus thermophilus	
Lactobacillus bulgaricus	Yoghurt
Lactobacillus helveticus	'Continental' cheeses, e.g. Emmenthal
Pediococcus cerevisiae	Fermented sausages
Staphylococcus carnosus	Fermented cucumbers
Lactobacillus plantarum	Fermented cucumbers and olives
Bifidobacterium bifidum	Fermented milk drinks

* The nomenclature of Garvie E.I. & Farrow J.A.E. (1982) *Int. J. Syst. Bact.* 32, 453–455.

only recently, however, that a systematic approach to production methods has been recommended. Thus Etchells *et al.* (Carr *et al.* 1975) have discussed the overall operation which should be adopted for the fermentation of cucumbers. A summary together with a rationale of their method is given below:

(a) *Grade* the cucumbers; diseased or damaged material is discarded

This minimizes the amount of pectinolytic enzymes in the cucumbers used for fermentation.

(b) *Wash* the cucumbers and immerse in brine (6.6% NaCl) containing chlorine (80 p/10^6)

Reduction in the level of microbial contamination.

(c) *Acidify* the brine and *supplement* it with acetate

Acetate under acid conditions is recognized to be a strong elective agent for *Lactobacillus* spp.

(d) *Inoculate* the brine with *Pediococcus cerevisiae* and *Lactobacillus plantarum*

The principal microorganisms in the 'natural' fermentation.

(e) *Sparge* the brine with nitrogen

Removes CO_2 and prevents the formation of 'bloaters'.

Subsequent studies have shown that the efficacy of (e) is dependent upon the size of the bubbles produced by the sparger; small bubbles are more effective than large ones in removing dissolved CO_2 from the brine. A recent observation that small 'buttons' of mould growth occur in sparged brines suggests that the oxygen content of the nitrogen used in sparging needs to be controlled otherwise the method used to ameliorate one fault will cause another.

(2) Fermented milk and cheese

It has long been recognized that the fermentation of lactose to lactic acid is a feature common to most if not all methods used in the production of fermented milks and cheese. Lactic acid bacteria, particularly species of *Streptococcus* (Table 5.4), are the organisms used most commonly in the fermentation of milk but *Bifidobacterium* spp. (Fig. 5.6) are used in certain fermented milk products in Japan. There has been a tendency for this phase of production to have been overemphasized perhaps because the extent and rate of change of the lactose content of the milk, and the relative ease with which these can be determined, contrasts with those associated with the milk proteins and fats. With both of the last mentioned, slow changes persisting long after the initial phase of lactose fermentation can have a profound effect on the flavour and texture of particular cheeses. With some cheeses, desirable flavours arise from the production of diacetyl from citrate by the starter culture (Fig. 5.7).

In addition to its role as a preservative agent and an inhibitor of certain food-poisoning organisms (e.g. *Staphylococcus aureus*), lactic acid contributes to:

(a) the curdling of the milk;

(b) the drainage of whey from the curd (it enhances syneresis of the coagulated proteins);

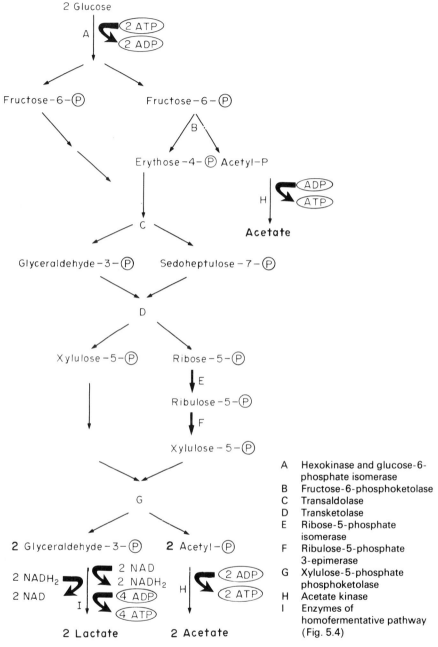

Fig. 5.6. A fermented milk drink is produced in Japan by fermentations based on *Bifidobacterium bifidum*. As will be noted from the pathway given above, this organism produces acetate and lactate (in bold) from glucose.

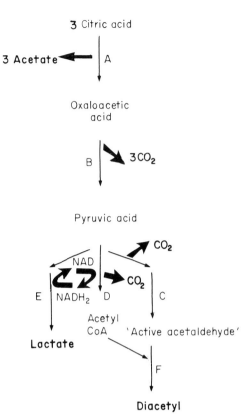

	A Citrate lyase
	B Oxaloacetate decarboxylase
	C An uncharacterized enzyme
	D Pyruvate dehydrogenase complex
	E Lactate dehydrogenase
	F Diacetyl synthase

Fig. 5.7. Diacetyl production from citrate by *Streptococcus lactis* subsp. *diacetilactis* gives cottage cheese a buttery flavour. Carbon dioxide is another principal product of the fermentation. As this gas can damage the texture of the curd and give it buoyancy, the fermentation of lactose in milk to lactic acid (Fig. 5.4) is done in one process and the harvested curd treated with cream in which citrate has been fermented to diacetyl. Plasmids are involved in coding for diacetyl fermentation.

(c) texture;
(d) elasticity;
(e) flavour of cheeses.

With (e) the contribution may be direct or indirect in the sense that the lactic acid establishes a pH that influences the action of enzymes, of milk or microbial origin, involved in the development of flavour through the modification of proteins, fats or products thereof. As a preservative, lactic acid functions through lowering the pH (4.5–5.3) thereby enhancing the antimicrobial action of salt and the reduced a_w resulting from the evaporation of water. As with the plant materials discussed above, the persistence of this contribution will depend upon the conservation of lactic acid. A low redox will ensure conservation not only of lactic acid but of oxygen-labile contributors to flavour such as H_2S and methanethiol in Cheddar cheese. It is reasonable to assume also that under farmhouse conditions, acetic acid would be produced during fermentation and that it would contribute to preservation. Claims have been made also that other metabolic products, some simple (H_2O_2) and others (e.g. lactobrevin) which remain poorly defined because they appear to have attracted the attention of etymologists rather than organic chemists, contribute to the preservation of cheese. Lactic acid can also act as an elective agent in the initiation of a particular microbial association or as a substrate for organisms that produce desirable flavours and changes in the physical appearances of cheese, such as Emmenthal. In this example, a product of lactic acid fermentation, propionic acid, can accentuate the antimicrobial properties of a cheese, especially against mould growth when the cheese is cut in the shop or home (Fig. 5.8).

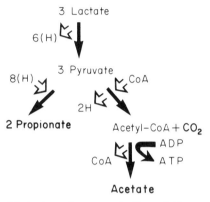

Fig. 5.8. In the ripening of Swiss-type cheese, propionibacteria ferment lactate to propionate (which imparts flavour and confers protection against mould growth on the surface of cut cheese), CO_2 (which produces the 'eyes' that characterize this type of cheese) and acetate.

Changes in the physical and chemical properties of the milk protein, casein, occur concomitantly with lactose fermentation. The addition of rennet (an enzyme isolated from the stomach of calves), either alone or in combination with pepsin, or proteases derived from *Mucor pusillus* or *Mucor milchii* bring about major changes. The rate of change can be accentuated if rennet is added to milk which contains thermostable proteases of microbial origin. With rennet, for example, peptide bonds between a phenylalanyl and a methionyl residue in K-casein causes destabilization of the casein micelle, the casein molecules aggregate and precipitate to form the curd. Further action by these enzymes will affect the taste and texture of a ripened cheese. As noted previously, microbial proteases remaining in pasteurized milk can produce bitter peptides during storage of cheese. Likewise, contamination of ripening cheeses with Group D streptococci can lead to the production of amines—tyramine, histamine and tryptamine—at above-average concentrations. As tyramine has been considered to be a possible initiator of migraines in susceptible subjects and produces acute symptoms in patients receiving monoamine oxidase inhibitors, the dairy microbiologist must strive to control contamination of the starting materials and thereby minimize undesirable changes in the nitrogenous materials of the cheese. Of course the nutritionally demanding lactic acid bacteria used for lactose fermentation will require a range of amino acids in order to grow. As milk does not contain sufficient free amino acids, the organisms have to satisfy their demands from the milk proteins, or be associated with organisms that can supply their needs. With the Group N streptococci, proteinases in the cell wall act on casein and the released peptides of 4–7 amino acid residues are taken into the cell. Cell-bound peptidase in some Group N streptococci (Fig. 5.9) break down peptides and the individual amino acids are taken into the cell. There is evidence that the rate of growth and acid production by *Streptococcus thermophilus* in milk is aided by the action of proteinase present in the cell wall of an associated starter culture, *Lactobacillus bulgaricus*. As the proteinases and peptidases of the starter cultures will be active long after the organisms have died, they contribute to flavour production in the stored cheese by releasing free amino acids, some of which may be precursors of desirable flavour components.

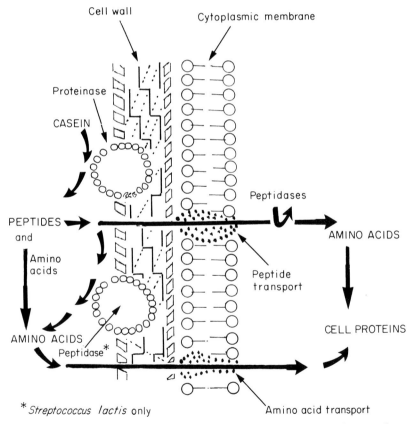

Fig. 5.9. The group N (lactic) streptococci have to satisfy their nitrogen requirements from the casein of milk. To achieve this, they have cell-bound proteinases and, in the case of *Streptococcus lactis*, cell-bound peptidases as well as peptide and amino acid transport systems. Plasmids are probably involved in the coding for proteinase activity.

STEPS IN PRODUCTION OF CHEESE

Many varieties of cheese undergo a 'ripening' process during which changes in appearance, texture, flavour and colour are brought about by organisms other than those associated with the initial fermentation of lactose. The ripening organisms (Table 5.5) may be within the cheese or confined to its

Table 5.5. Organisms associated with the ripening of cheeses.

Organism	Example of cheese
Lactobacillus casei	Cheddar cheese
Propionibacterium spp.	Swiss cheeses
Penicillium roqueforti	Blue cheese, e.g. Roquefort
Penicillium camamberti	Camembert
Brevibacterium linens and yeasts	Brick cheese

surface ('smear ripened'). It can be surmised that our forefathers must have experienced many failures in their attempts to select shape, size, texture and methods of production before they arrived at those which were optimal for the growth of particular ripening organisms and hence those changes in texture, flavour and appearance sought by a discerning consumer.

The steps in the production of cheese can be summarized thus:

1 *inoculating* the milk with starter cultures (Table 5.4);
2 *coagulating* (curdling) the milk proteins by addition of enzymes such as rennet;
3 *removing* the whey and giving the curd a texture and a form or shape;
4 *adding* salt to aid flavour, texture or keeping quality;
5 *providing* conditions and organisms essential for ripening.

Types of cheese

Through variations of all these stages or by the omission of some, products conforming to one or other of the following broad groupings can be achieved.

(a) Fermented milks

Yoghurt, Kefir, Koumiss

Milk is fermented with starter organisms, bacteria and/or yeasts, which contribute to curdling and flavour. The curd is *not drained*. The product has a short storage life

(b) Unripened cheese

Cottage cheese

Milk is fermented with starter organisms which contribute to curdling and flavour.

The curd is *cut*, heated, and the whey allowed to *drain*. The curd is washed in chilled water to which NaCl may be added.

The milled curd has a moisture content of 60–80% and a storage life at chill temperature of *c.* two weeks

(c) Ripened cheese

    (i)   Soft

        Camembert

The milk is fermented with starter organisms which contribute, along with rennet, to curdling.

The curd is put into perforated containers and the whey allowed to *drain* and a disc of about 5 × 1 inch formed. The surface of the consolidated curd is treated with salt and colonized by ripening organisms. The moisture content is *c.* 40–50%.

    (ii)   Semi-soft

        Roquefort, Blue cheese, Gorgonzola, Limburger

Milk is fermented with starter organisms which contribute along with rennet to curdling.

The curd is *cut* and the whey allowed to *drain*; it is salted and inoculated with spores of *Penicillium roqueforti* (with Limburger, the spores are put on the surface of the cheese).

The curd is put into metal hoops and the whey allowed to *drain*. The surface of the cheese is salted, covered with paraffin wax to prevent smear ripening and punctured with needles to encourage the diffusion of O_2 and CO_2 and thus the growth of the moulds within the cheese. Moisture content of cheese, *c.* 45%.

    (iii) Hard

        Cheddar

Milk is fermented with starter organisms which contribute along with rennet to curdling.

The curd is *cut* and the whey allowed to *drain*; it is broken down to particles ($\frac{1}{2} \times \frac{1}{2} \times 1$ inch), salted and packed in hoops to which pressure is applied to *force* out whey.

The cheese is taken from the hoop, its surface allowed to dry before being waxed and stored.

The cheese has a moisture content of *c.* 35% and a long storage life.

        Swiss cheese

Propionibacterium spp. break down the lactic acid to propionic acid, acetic acid (flavour) and CO_2 (the 'eyes' or gas bubbles which are a characteristic feature of these cheeses (Fig. 5.8)).

    (iv) Hard, grating cheese

        Parmesan

Production methods similar to those used for hard cheese; the grating cheeses are stored for a long time and a low moisture content (*c.* 32% moisture) is achieved. Such cheeses have a long storage life.

Starter cultures

The lactic acid bacteria which are used as starter cultures are listed in Table 5.4. Traditionally the choice of organism(s) would have been determined by:
1 the temperature at which the milk was held before and during the lactose fermentation phase;
2 whether or not a characteristic flavour was sought.
Thus the yoghurt, cultures of *Streptococcus thermophilus* and *Lactobacillus bulgaricus* are added to milk at 48 °C and incubated at 42–45 °C, temperatures at which both organisms grow and produce diacetyl (the streptococcus) and acetaldehyde (the lactobacillus). These two substances contribute the characteristic flavour of yoghurt. With milk held at lower temperatures throughout production, for example 30–39 °C with Cheddar cheese, the mesophilic lactic (Group N) streptococci, *Streptococcus cremoris* and/or *Str. lactis*, are used. With hard cheeses of the Swiss-type, the temperature of the curd and milk spans the range 31–54 °C and this calls for thermoduric cultures, *Str. thermophilus* and *Lactobacillus helveticus*.

The ripening organisms

The organisms associated with the ripening of cheese are listed in Table 5.5. As noted above in some cheeses, e.g. Cheddar, the starter cultures can be the principal contributors to ripening. The ripening organisms may be included with the starter cultures, as with *Propionibacterium* spp., in the cultures used in the production of Swiss cheese. With blue cheeses, the spores of *Penicillium roqueforti*, harvested from moulds growing on bread crumbs, may be added to the milk or the curd; spores of *Penicillium camemberti* are added to the surface of the formed cheese. The colonization of the surface of cheeses which are smear ripened often depends upon the transfer of organisms from the general environment of the cheese factory and the establishment of a microbial succession, initiated by acid-tolerant film yeasts, which grow at the expense of the lactic acid, and climaxed by *Brevibacterium linens*. The initial election of organisms is determined by the salt and acid at the surface of the cheese; as the lactic acid is used by the pioneering yeasts the conditions become selective for the salt-tolerant *B. linens*. The latter, a coryneform organism, is actively proteolytic. Some strains of this group of organisms are induced to synthesis pigments when exposed to light, a trait which may contribute to the colour of the cheese. The elective action of salt is assured throughout ripening by the practice of washing the cheeses in brine; the frequency and thoroughness of this operation determines the amount of microbial growth and hence the extent of ripening.

Substrates used by ripening organisms

There are three main substrates available to the organisms which are associated with ripening:
(a)   lactic acid;
(b)   fats;
(c)   proteins.

The lactic acid may be fermented or oxidized. It is fermented (Fig. 5.8) in Swiss cheese by *Propionibacterium* spp. and the end-products of fermentation contribute to the flavour (propionic and acetic acids), appearance (the CO_2 forms bubbles or 'eyes'), and microbial stability of the cheese. It is oxidized by the film yeasts which colonize the surface of cheese and in consequence the pH drifts back to a neutral reaction thus allowing the growth of the less acid-tolerant *Brevibacterium linens*. The drift in the pH (4.5–6.0) of blue cheese is associated with the oxidation of lactic acid by *Penicillium roqueforti*. The temperature and humidity of the stores at the outset of ripening of Roquefort-type cheeses are selected so that microbial growth is encouraged; thereafter storage is at lower temperatures in order that the enzymes produced by the fungus can bring about ripening without luxuriant mould growth occurring. A succession of non-acid-tolerant organisms in such cheeses is probably prevented by the levels of O_2 (low) and CO_2 (high) achieved because of the relatively feeble diffusion of these gases along the tracks remaining after the cheese has been pierced with long wires. Subsequent to the breakdown of lactic acid, *P. roqueforti* produces lipases which lead initially to the accumulation of acetic, butyric, caproic caprylic and capric acid. Subsequently methyl ketones, acetone, 2-butanone, 2-pentanone, 2-hexanone, 2-heptanone and 2-onanone, accumulate and it is considered that these, especially 2-heptanone, are the principal flavour compounds of Roquefort. With cheese such as *Limburger*, the change in texture and flavour is associated with the digestion of casein by *B. linens*.

Properties

It was noted at the beginning of this chapter that fermentation is a very ancient method of preparing or preserving foods. It can be assumed that the development of successful methods was the outcome of much trial and error. Indeed as most if not all of the traditional methods can be likened to the enrichment cultures used in contemporary microbiology, our forefathers based the selection of methods on the phenotypic properties of agents, microorganisms, of whose existence they were unaware. The range of phenotypic properties selected for food production over thousands of years is reflected in the information given in Table 5.2. Although the application of microbiology to studies of processes such as cheese-making demonstrated that lactic acid bacteria were of cardinal importance, a detailed understanding of these organisms' contribution had to await the studies of biochemists and molecular biologists. Until recently dairy microbiologists would have claimed that a successful starter culture of Group N streptococci,

Streptococcus cremoris and *Str. lactis*, possessed three principal phenotypic attributes:

(a) a proteinase system with which they satisfied their nitrogen requirements at the expense of casein;

(b) a lactose fermentation system that resulted in the rapid production of lactic acid;

(c) a resistance to the lactoperoxidase system in milk (Fig. 5.11).

Recent studies have demonstrated that lactic streptococci have an exceptionally large complement of plasmid DNA, upwards of 14 plasmids have been found in a single strain. Although only a few plasmids have been shown to code for a particular physiological attribute of an organism, those few have been clearly linked with the success or otherwise of starter cultures under commercial conditions; for instance, plasmids code for lactose metabolism (Fig. 5.10), the production of proteinase enzymes (Fig. 5.9), the fermentation of citrate (Fig. 5.7), resistance to inorganic salts and nisin production.

The broad outline of the steps leading to lactic acid production from lactose is shown in Fig. 5.10. In *Streptococcus thermophilus*, the lactose is transported across the cell wall by a permease whereas with Group N streptococci, lactose is taken up as glucosyl-$\beta(1,4)$-galactoside-6-phosphate via a phosphoenolpyruvate-dependent phosphotransferase system. A recent survey of *Str. lactis* from many sources has shown that those which had been selected as starter cultures because of their rapid production of lactic acid hydrolyse lactose with β-phosphogalactosidase, which is plasmid coded, whereas the non-dairy strains, which ferment lactose slowly and produce various end-products, have both β-galactosidase and β-phosphogalactosidase. Although knowledge about the physiology of the starter cultures aids an understanding of cheese-making, the successful production of cheese on a day-to-day basis calls for a microbiologist to provide the cheese maker with cultures having the three cardinal attributes noted above. An outline of the methods which have evolved to meet the requirements of commercial cheese production is discussed in the following section.

Production

Unlike the industries which produce fermented plant materials, the dairy industry has for many years been using pure cultures of bacteria to inoculate the milk used for cheese production. As the scale of cheese-making equipment increased and as the process developed from a craft into a technology, so the demands for cultures which perform in a predictable manner increased. Until recently, the production of starter cultures was based on the techniques of batch culture and they were produced in a laboratory at the cheese factory. The microbiologist had three main objectives:

1   the selection and maintenance of microorganisms having properties appropriate to the methods used for cheese production;

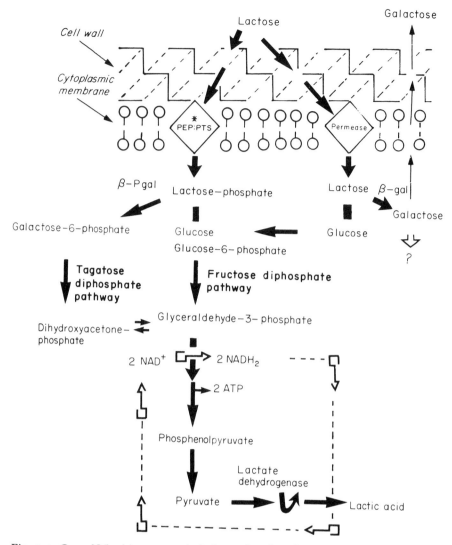

Fig. 5.10. Group N (lactic) streptococci take lactose into the cell as glucosyl-β-(1,4)-galactoside-6-phosphate via a phosphoenolpyruvate-dependent phosphotransferase system (PEP:PTS). Plasmids are probably involved in one or several steps of this system. With *Streptococcus thermophilus*, it is likely that a permease is involved in the transport of the milk sugar, lactose, across the cytoplasmic membrane.

2 the control of contamination of starter cultures with yeasts, coliforms, moulds and bacteriophages;

3 the provision of adequate quantities of culture which are able to produce lactic acid at a rate and in the quantity required by the factory schedule. In hard cheese production, the inoculum of starter cultures is 1–2% (w/v) of the milk to be fermented.

The routine examination of starter cultures for contamination with extraneous organisms can be based on a combination of indirect and direct methods. As the starter cultures should contain only catalase-negative lactic acid bacteria, there ought not to be any evolution of O_2 when H_2O_2 is added to a culture. Likewise, as the lactic acid bacteria do not produce H_2 from the fermentation of carbohydrates, the casein clot formed in the growth medium, either heat-treated milk or a medium of antibiotic-free skim milk (12% w/v), should not be fissured by gas. Such fissures are seen in cultures contaminated with coliform organisms and it is for this reason that glass containers have been used traditionally for the propagation of starter cultures. Such containers also permit the experience microbiologist to check routinely the form and quality of the clot formed by the starter organisms in media containing milk. Milk containing bromocresol purple can be used to demonstrate indirectly whether or not phages are contaminating a starter culture. If phages are present, then the lactic acid bacteria will not grow in the BCP milk and the pH indicator will remain blue in colour.

With the direct methods of examining for contamination, the following tests can be applied.

(a) A starter culture is streaked out on Malt Agar—any yeast or mould contaminants will form colonies during incubation at 26 °C for 3–5 days.

(b) A loopful of starter culture is used to inoculate MacConkey broth—if acid and gas are formed during incubation at 30 °C for 3 days, then it is presumed that contamination with coliforms has occurred.

(c) A filtrate of starter culture is applied to an agar medium containing an inoculum of a starter culture—the formation of plaques will be evidence of phage infection of the starter culture.

Two tests can be used to assess whether or not a culture of a lactic acid bacterium has phenotypic properties appropriate to a particular schedule for cheese production. With the *activity test,* pasteurized milk is inoculated with the organism and the amount of acid formed during incubation at 30 °C for 6 h determined by titration with NaOH and phenolphthalein as indicator. With the *vitality test,* a laboratory method which stimulates cheese production is used. Milk inoculated with the culture is mixed with rennet and titration with NaOH and phenolphthalein used to measure acid production during incubation at 30 °C of the whole milk and then in the curds recovered in a sieve.

When strains of lactic acid bacteria having properties appropriate to the cheese-making schedule have been obtained in pure culture, their day-to-day management has to be based on a regime whereby a 'physiologically fit' organism is given to the cheese maker. The latter will experience difficulties in production if supplied with a 'slow' culture. This can be caused by incompetent control in the laboratory. Thus the use of an unsuitable medium, especially milk that has received drastic heat treatment, together with irregular sub-culturing can select organisms which are: (i) nutritionally demanding; (ii) intolerant of acid conditions even though they bring about a rapid initial fermentation of lactose, or (iii) tolerant of very acid conditions but they ferment lactose slowly.

Slowness may be caused—and is probably most often caused in large cheese factories—by phage infection of the starter culture. As the micro-biologist may be called upon to produce several hundred gallons of starter cultures per day, there will be a need to have a programme whereby cultures, beginning with freeze-dried cultures or cultures in test tubes, are propagated in larger and larger volumes of milk until an inoculum of sufficient size is available for a container holding upwards of 500 gallons of milk. As air is probably the most important vector in contamination of cultures with phage, the problem facing the dairy microbiologist can be appreciated. Several policies have been adopted to overcome this problem.

1 As the phages are strain specific, several strains of starter culture are used in combination (Table 5.6) or in rotation so that a build-up in the level of contamination of the factory with a particular phage is minimized.

2 Attempts are made to select phage-resistant strains of starter cultures.

3 The starter cultures are grown in a medium depleted of Ca^{2+} by addition of phosphates so that the phage cannot absorb on to the host.

4 The starter culture is 'protected' at all stages of preparation from contamination with phage. This, the Lewis method, calls for specially designed equipment whereby inoculation ports in large containers, often pressurized, are fitted with seals containing chlorine at $250\,p/10^6$ (see Fig. 3.2).

Table 5.6. Composition of starter cultures used for hard cheese. (From Cox, Stanley & Lewis (1978) in *Streptococci* (Eds Skinner F.A. & Quesnel L.B.). Academic Press, London.)*

Type	*Streptococcus* species	Method of use and characteristics	Cheese variety/ location
Single-strain starters	*Str. cremoris* *Str. lactis* *Str. lactis* subsp. *diacetilactis*	Single or paired	Cheddar in New Zealand and Australia and selected UK creameries
Multiple-strain starters	*Str. cremoris* *Str. lactis* *Str. lactis* subsp. *diacetilactis* *Leuconostoc* spp.	Defined mixtures of two or more strains (may be used in pairs)	Hard cheese varieties in US and certain UK creameries
Mixed-strain starters	*Str. cremoris*	Unknown proportions of different strains which can vary on subculture (may be used in pairs)	Most hard cheese varieties in UK and Europe

* See Table 5.4 for revised nomenclature.

It was implied above that there are changes occurring in the dairy industry and over the past few years much effort has been devoted to a marriage of the technology of the fermentation industries and the needs of the cheese maker. Thus it is now possible to grow starter cultures in large fermenters having critical control of temperature, pH and nutrient status, to concentrate the cells by centrifugation and to store and distribute the cultures at temperatures as low as that given by liquid nitrogen. As with the production of Bakers' yeast, this means that production of starter cultures can be done on a large scale, remote from the place of use.

A slow fermentation of lactose need not be the fault of the culture supplied to the cheese maker, it can be due to the milk being an un-favourable medium; for example, antibodies (agglutinins) in the milk can agglutinate starter cultures, a problem that is of particular importance when rennet is not used and the antibodies remain in suspension. Slow fermenta-tion may be due to excessive contamination of the milk—the starter culture cannot compete with the resident flora—or the production of an antibiotic by an organism related to the starter culture. Thus nisin, a polypeptide produced by *Streptococcus lactis*, inhibits *Str. cremoris*. As the heat treatment of the milk media used to propagate starter culture inactivates the lacto-peroxidase system (Fig. 5.11), lactic acid bacteria sensitive to this system can

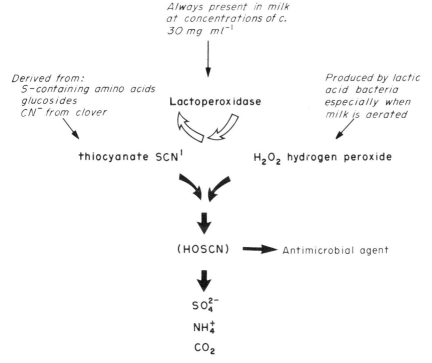

Fig. 5.11. Although mammalian milk is recognized as being a rich source of nutrients, the growth of microorganisms is impeded by a number of antimicrobial systems. Of these, the lactoperoxidase system is of particular interest.

be selected and these may fail to initiate fermentation in milk intended for cheese production. Indeed the day-to-day management of the mother cultures ought to include tests that identify the lactoperoxidase-sensitive strains. The milk may contain inhibiting concentrations of those antibiotics which have been used therapeutically to treat mastitis, or disinfectants used on milking equipment. On a well-ordered farm milk from cows receiving antibiotics will not be offered to the dairy. As this practice cannot be taken for granted, tests have been devised for the rapid antibiotic assay of milk received by a dairy. The assay, which is based on lactic acid production by *Str. thermophilus* with bromocresol purple as indicator, will detect 0.02 iu of penicillin ml^{-1}. These are but a few of the factors which have been associated with the failure of starter cultures and they draw attention again to the need for a microbiologist to consider all the stages in the production of a raw material. With milk production in the UK, the emphasis throughout all stages is placed on asepsis, keeping out organisms during milking—by the thorough washing of the udders and the use of sterilized equipment—the cooling of freshly drawn milk, and the maintenance of low temperatures are of vital importance during its collection and storage. The milk used for cheese production is normally pasteurized; not only does this reduce the size of the resident flora but it causes chemical changes which make milk a more favourable medium for the growth of starter organisms.

6 Microbial Food Spoilage

A study of the literature on food microbiology can leave the impression that food microbiologists have been concerned more with failure than with the mechanisms contributing to successful preservation. Thus a cursory survey of the literature permits a listing of upwards of 50 of the approved genera of bacteria that have been associated with (1) food spoilage, (2) food poisoning, (3) food-borne disease, or (4) general contamination. An equally large number of genera of fungi and yeasts could be assembled for categories (1) and (4). Since the review by Mossel and Ingram (1955) attention has been given to the definition of the association of micro-organisms developing on, rather than a cataloguing of all the organisms recovered from spoiled foods. Such an approach allows general statements to be made, viz. Gram-negative, non-aciduric, nutritionally non-fastidious bacteria are commonly associated with the spoilage of moist proteinaceous foods, related aciduric genera with acid fruit products, lactic acid bacteria with foods rich in nutrients and fermentable carbohydrates, and fungi with plant materials. Additional examples are given in Table 6.1.

With the first-mentioned category, the gaseous environment, temperature and the addition of sodium chloride, nitrite or glucose to meat can influence profoundly the composition of the microbial association (Table 1.2). It must be recognized that the use of the concept of a microbiological association by food microbiologists has a utilitarian rather than a fundamental significance simply because it is applied to the consortium of organisms present at the time that a food is deemed to be unacceptable for human use. Thus the association may, and often does, represent only the stage at which the pioneering organisms of what would be a succession under 'normal' conditions have achieved or are approaching climax. Indeed, though this initial climax is associated with macroscopic and/or chemical changes of such a magnitude that a food is rejected (Fig. 6.1), the bulk of the food need not have changed to any appreciable extent. Observations of the changes occurring in milk provide a good example of this situation. Milk would be regarded as sour should it curdle on addition to a hot beverage or if it had gelled in the container. These manifestations of spoilage are due almost entirely to microbial fermentation of lactose and curdling results from the pH dropping below the pH (5.4) of casein. With prolonged storage, a succession of organisms will use lactate arising from lactose fermentation,

Table 6.1. A broad grouping of microbial associations and spoiled foods.

General characteristic of food	Examples of food	Microorganisms
A. 'Moist' foods		
Rich in proteins, vitamins but deficient in fermentable carbohydrates	Meat, fish, eggs and poultry	*Pseudomonas, Acinetobacter, Alternomonas, Micrococcus, Aeromonas, Proteus*
Rich in proteins, vitamins and fermentable carbohydrate	Milk	*Streptococcus, Lactobacillus, Pseudomonas, Microbacterium lacticum*
Rich in fermentable carbohydrates and H-ions	Fruit and plant juices	*Streptococcus, Lactobacillus, Pediococcus,* Acetic acid bacteria, yeasts and moulds
Rich in proteins and vitamins but deficient in fermentable carbohydrates, salt and NO_2^- added	Bacon and ham	*Micrococcus, Lactobacillus, Vibrio, Streptococcus,* yeasts and moulds
Rich in carbohydrates, water and H-ions	Fruits	*Lactobacillus, Acetobacter, Rhizopus, Penicillium, Botrytis, Kloeckera, Hanseniaspora, Pichia, Torulopsis*
A general range of nutrients, a tendency to dry-out during storage; biological organization including an integument part of which may be damaged during harvesting and grading	Vegetables such as lettuce, legumes, carrots and brassicas	*Erwinia, Pseudomonas, Sclerotinia, Rhizopus, Fusarium, Phytophthora, Penicillium, Phoma, Pythium, Peronospora, Alternaria*
Rich in proteins and carbohydrates; acetic acid as an ingredient	Mayonnaise-type sauces and coleslaw	*Lactobacillus, Saccharomyces*
B. 'Dry' Foods		
Rich in carbohydrates	Cereals, flour	*Aspergillus, Fusarium, Monilia, Rhizopus, Penicillium, Erwinia herbicola (Enterobacter Agglomerans)*

the casein and fats and thereby bring about mineralization of the milk; all changes after the curdling phase would be of academic interest only to a food technologist. A succession in the types of microorganisms and chemical changes can be observed when frozen peas are thawed and stored at room temperature (Fig. 6.1).

It is now recognized that the adoption of the term 'association' signalled an important change in food microbiology because it directed attention at the few microorganisms among the initial contaminants that brought about spoilage of particular foods. In practice, 'association' has connotations of interaction/interdependence between organisms in an environment. As food-spoilage microorganisms can be considered to be little more than opportunist

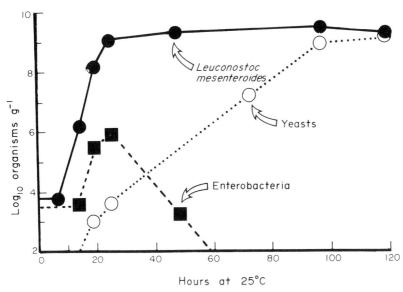

Fig. 6.1. When frozen peas are thawed and stored at room temperature their surfaces become clothed with a large population of dextran-forming *Leuconostoc mesenteroides*. The graph shows the growth of this organism and also the short period of growth of members of the Enterobacteriaceae and the subsequent emergence of yeasts as dominant organisms (David Kelly, unpublished observations). Some chemical changes occurring with these changes in the size of the microbial populations are shown in Fig. 6.2.

saprophytes, it may well be that in future the term 'association' will be replaced by 'bloom'. The latter is of common use in general microbiology to describe the temporal dominance of microorganisms in an environment. In the meantime, however, the term 'association' will be used in this chapter.

Studies of the composition of the microbial associations (Table 6.1) developing on foods of known intrinsic properties and stored under reasonably well-controlled conditions have lead to a better understanding of the implicit characteristics of spoilage organisms and, in some cases, the identification of key properties that fit an organism to a niche when it offers an environmental factor in an extreme form. Such studies show that in those foods that have no intrinsic property far removed from what is considered to be optical for the growth of a wide range of micro-organisms and a buffering system which prevents an extreme form developing, the spoilage organisms tend to be those, such as *Pseudomonas* spp., which appear to be well adapted for a scavenging/pioneering existence in the degradative stages of the carbon and nitrogen cycle. Such foods commonly harbour a diverse flora at the time of production and the success of the few species that form an association need not result in the death of the other contaminants. As was noted previously (Table 1.2), extrinsic factors such as the composition of the atmosphere and temperature can have a profound elective influence on the initial contaminants and ultimately it is probably the rate of growth that determines whether or not a contaminant

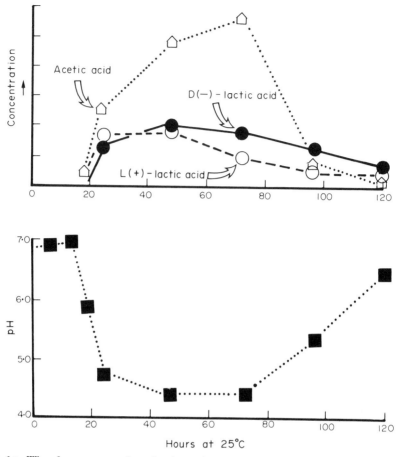

Fig. 6.2. When frozen peas are thawed and stored at room temperature, changes in the pH and organic acid content occur. The progressive increase in the concentration of acetate and lactate shown in the graph is associated with the growth phase of *Leuconostoc mesenteroides* (Fig. 6.1). It is probable, moreover, that the rapid decline of Enterobacteriaceae is associated with the accumulation of acetate and a declining pH. The ultimate collapse in the concentration of acetate is associated with the latter part of the growth phase of yeasts (David Kelly, unpublished observations).

contributes to the association present at the time of spoilage of foods such as meat. With milk the poor buffering capacity does not prevent a rapid drift in pH resulting from lactose fermentation and conditions obtain which select organisms tolerant of an acid environment, viz. the lactic acid bacteria. The many roles played by these organisms in the food industry (Table 6.3) highlight the problems faced by anyone who attempts the definition of a food-spoilage organism. Indeed as with weeds, such an organism is, in practice, the wrong organism in the wrong place at the wrong time.

The methods and ingredients used in the production of certain foods lead to the enrichment of microorganisms that become well-known to the food microbiologist but of concern only to a relatively few general microbiologists. Such an organism is *Brochothrix (Microbacterium) thermosphacta*.

Table 6.2. Bacteria associated with spoiled foods.

Implicit properties of organisms	Organisms	A			B	C	D	E	F	G	H	I
		1	2	3								
Gram-negative aerobes, facultative anaerobes. Little to moderate acid tolerance. Relatively simple nutrients support growth	Pseudomonas	+	+	+	+							
	Alcaligenes	+	−	−	+							
	Enterobacter	+	−	−	+						+	
	Proteus	+	−	−	−	(+)						
	Vibrio	−	−	−	−	+						
	Aeromonas	+	−	−	(+)	−						
	Acinetobacter	(+)	+	+	−	−						
Gram-positive, aerobic, faculative anaerobic cocci. Marked tolerance of salt; moderate tolerance of acid. Some strains thermoduric	Micrococcus			+	+	+					+	
	Staphylococcus				+	+						
Gram-positive rods and cocci; fermentative metabolism; moderate to marked tolerance of acid. Some strains thermoduric. Complex nutrients required	Streptococcus				+	(+)	+	−	−	−	+	
	Leuconostocs				−	−	+	+	−	−		
	Pediococcus				−	−	+	+	−	−		
	Aerococcus				−	−	−	−	−	−		
	Lactobacillus				+	+	+	+	−	+		
Gram-negative, acid-tolerant rods. Obligate aerobes	Gluconobacter				−	−	+	−				
	Acetobacter				−	−	+	+				
	Frateuria				−	−	+	−				

Key to classes of food

A Proteinaceous foods with small content of fermentable carbohydrate
  1. Eggs—storage in range 1–25 °C
  2. Red meat stored at 0–4 °C
  3. Red meat stored at 10–15 °C
B Milk
C Processed meats (bacon, ham, etc.)
D Fruit juices—fresh or fermented
E Foods preserved with acetic acid
F Salted fish
G Vegetables stored in brine with SO_4^{2-} contamination
H Pasteurized milk
I Canned foods

(continued)

Table 6.2—*Continued*

Implicit properties of organisms	Organisms	Classes of food spoiled — A 1	A 2	A 3	B	C	D	E	F	G	H	I
Gram-negative, acid-tolerant fermentative rod	*Zymomonas*											
Anaerobic respiration with SO_4^{2-}. Spores formed by *Desulfotomaculum*	*Desulfovibrio*						+					
	Desulfotomaculum											+
Obligate requirement for high concentration of NaCl	*Halobacterium*								+	+		
Spores having marked resistance to heat	*Bacillus*										+	+
	Clostridium										+	+

Key to classes of food

A   Proteinaceous foods with small content of fermentable carbohydrate
   1. Eggs—storage in range $1–25\,°C$
   2. Red meat stored at $0–4\,°C$
   3. Red meat stored at $10–15\,°C$

B   Milk

C   Processed meats (bacon, ham, etc.)

D   Fruit juices—fresh or fermented

E   Foods preserved with acetic acid

F   Salted fish

G   Vegetables stored in brine with SO_4^{2-} contamination

H   Pasteurized milk

I   Canned foods

+ Principal, or (+) occasional contaminants.

128

Table 6.3. Lactic acid bacteria associated with foods.

Genus	Species	1	2	3	4	5	6	7	8	9	10	11	12	13	14	15
Streptococcus	*thermophilus* ***	ST	—	ST	—	—	—	—	—	—	—	—	—	—	—	—
	lactis	ST	ST	ST	—	—	—	—	—	—	—	—	—	—	—	—
	cremoris	ST	ST	ST	—	—	—	—	SP	—	—	—	—	—	—	—
	faecium	c	—	—	—	—	—	—	—	—	—	—	—	c	—	—
	faecalis	c	—	—	F	—	—	—	—	—	—	—	—	c	—	—
Leuconostoc	*mesenteroides*	c	—	—	F	—	F	SP	—	—	—	—	—	—	—	—
	dextranicum	ST	ST	ST	F	—	—	—	—	—	—	—	—	—	—	—
	lactis	c	ST	—	—	—	—	—	—	—	—	—	—	—	—	—
	cremoris	ST	ST	ST	—	—	—	—	—	—	—	—	—	—	—	—
	oenos†	—	—	—	—	c	—	—	—	—	—	—	—	—	—	—
Pediococcus	*cerevisiae*	—	—	—	F	SP	F	—	—	—	—	—	—	—	SP	—
	acidilactici	—	—	—	c	c	—	—	—	—	—	—	—	—	—	—
	pentosaceus	—	—	—	c	c	—	—	—	—	—	—	—	—	—	—
	halophilus	—	—	—	—	—	—	—	—	R	—	—	—	—	—	—
	urinae-equi	—	—	—	—	c	—	—	—	—	—	—	—	—	—	—
Aerococcus	*viridans*‡	—	—	—	—	—	—	—	—	—	—	PA	—	c	—	—

Key to Foods

1 Cheese
2 Butter
3 Yoghurt, koumiss, kefir
4 Fermented vegetables
5 Fermented fruit juices and beverages
6 Fermented sausages
7 Sugar cane and syrups
8 Cured meats
9 Soy sauce
10 Sour-dough bread
11 Pathogen of lobster

Key to Symbols

ST Starter culture
c Casual contaminant
R Ripening organism
F Contributes to fermentation
SP Spoilage organism
PA Pathogen of lobster
— Not common contaminant
* See Table 5.4 for revised nomenclature.

(continued)

Table 6.3—*Continued*

Genus	Species	Foods 1	2	3	4	5	6	7	8	9	10	11	12	13	14	15	Key to Foods	Key to Symbols
Lactobacillus	*brevis*	c	c	c	F	SP	—	—	—	—	F	—	SP	—	SP	—	12 Foods preserved with acetic acid	** *Str. raffinolactis* is a common contaminant of raw milk (Garvie, 1978).
	buchneri	—	—	—	—	SP	—	—	—	—	—	—	SP	—	SP	—		
	fermentum	c	c	c	—	SP	—	—	—	—	F	—	—	—	SP	—		
	plantarum	—	—	—	F	c	—	—	—	—	F	—	SP	—	SP	F	13 Curing brines	
	viridescens	—	—	—	—	—	—	—	SP	—	—	—	—	—	—	—	14 Marinated herring	
	bulgaricus	ST	—	ST	—	—	—	—	—	—	—	—	—	—	—	—		† This is probably a species of a genus other than *Leuconostoc*
	delbruecki	—	—	—	—	—	—	—	—	—	F	—	—	—	—	—	15 Preparation of coffee and cocoa beans	
	collinoides	—	—	—	—	c	—	—	—	—	—	—	—	—	—	—		
	fructivorans	—	—	—	—	—	—	—	—	—	—	—	SP	—	—	—		
	casei	c	c	c	—	—	—	—	—	—	F	—	SP	—	—	—		‡ Synonym, *Gaffkya homari*
Atypical streptobacteria		—	—	—	—	—	—	—	SP	—	—	—	—	c	—	—		

It was isolated originally (1951) from American pork sausages and it is now recognized as the dominant bacterial contaminant of British fresh sausages. Although it is a general contaminant of red meats, no one has identified its niche in nature. A similar situation obtains with *Kurthia* spp., casual contaminants of red meats and areas in which meat is processed. These organisms are notable for the feather-like colonies formed on the surface of nutrient agar (Fig. 6.3) and the growth cycle in which short rods grow into long rods associated in long chains, some of which twist on themselves (Fig. 6.4). Short rods are formed at the end of the phase of active growth.

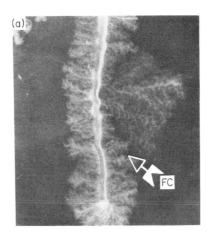

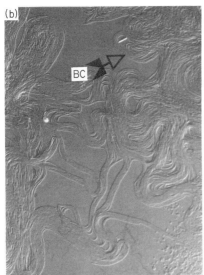
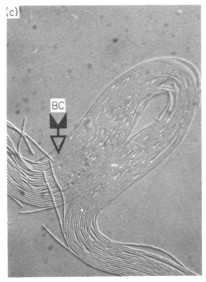

Fig. 6.3. *Kurthia zopfii*, a casual contaminant of meat and meat processing equipment, forms a feather-like colony (FC) on the surface of nutrient agar (a). Examination of the growth on nutrient agar (b and c) with the light microscopes reveals long chains of cells in bundles (BC) which snake-out over the surface of the medium (Dee Gascoyne, unpublished observations).

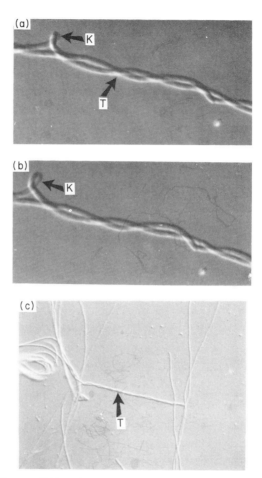

Fig. 6.4. Twisted filaments [T in (a)] are common features in the feather-like growth of *Kurthia zopfii* on the surface of nutrient agar (Fig. 6.3). The origin of such a pair of twisted filaments (K) is evident in (a) and (b) and the final form in (c) (Dee Gascoyne, unpublished observations).

Thus these organisms have certain features in common with *Br. thermosphacta*; the former differs from the latter in being an obligate aerobe, an active producer of catalase and motile (Fig. 6.5).

It was noted above and in Chapter 1 that, in general, a diverse flora is a common feature of foods offering little in the way of extremes of environmental gradients. When extreme environmental factors obtain, either for a short period or throughout the life of the food, then selection favours organisms that have particular physiological or morphological adaptations. The selective influences of temperature and water activity are shown in Fig. 6.6. In practice, the more extreme such elective factors become, the smaller the range of microorganisms having physiological adaptations that permit them to colonize a food. In other words, the food acts as an enrichment medium. The physiological adaptations that fit an organism to an environment rich in salt or syrups were discussed in

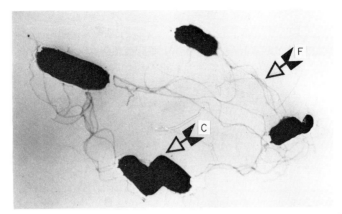

Fig. 6.5. *Kurthia zopfii* is motile; flagella (F) are arranged around the entire surface of the cell (C).

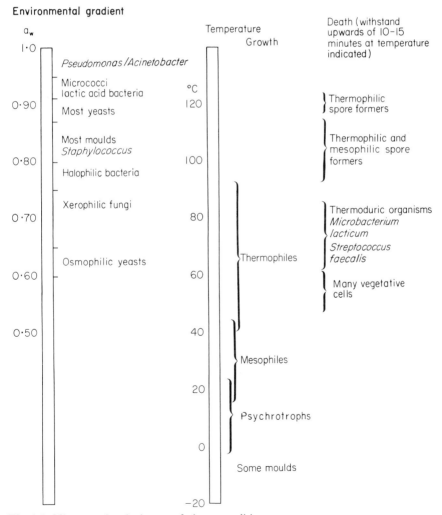

Fig. 6.6. Microorganisms' tolerance of adverse conditions.

Chapter 2. The presence of specific chemicals in a food also has an elective action. It was noted in Table 6.1 that an acid environment is elective for yeasts, lactic acid bacteria and aciduric Gram-negative bacteria such as *Acetobacter*, *Gluconobacter*, *Frateuria* and *Zymomonas* (see also Table 6.4).

Table 6.4. Gram-negative bacteria associated with fermented products.

Bacteria	Associated with
Aerobes	
Acetobacter	Acetification of fruit juices, wines and beers
Gluconobacter	Commercial production of vinegar
Frateuria	
Facultative anaerobes	
Hafnia protea	Minor defects in beer flavour and yeast
(*Obesumbacterium proteus*)	performance
Anaerobes (at least when first isolated)	
Zymomonas mobilis subsp. *pomaceae*	Cider sickness—unpleasant flavour due to accumulation of acetaldehyde
Zymomonas mobilis subsp. *mobilis*	Common in fermenting fruit juices in the tropics— viz. pulque, the fermenting sap of *Agave americana*
Anaerobe—strict	
Pectinatus cerevisiiphilus	Isolated from spoiled beer

If acetic acid is an important component of a food, then lactic acid bacteria and yeasts are enriched and form the spoilage associations in products such as mayonnaise and coleslaw (Table 6.1). When this acid is used as the principal preservative, then spoilage of a food can result from the growth of a few, very highly adapted microorganisms (Table 6.5). The list of organisms in Table 6.5 demonstrates that the extreme environments resulting from the preparation or preservation of particular foods have led to the isolation of microorganisms notable for their tolerance of some inimical feature. *Saccharomyces bailii* is the epitome of such an organism.

The application of temperatures above that of the surroundings to foods also has an elective action, either by selecting organisms capable of growth at elevated temperatures or survival at relatively high temperatures. In the latter instance, resistance may be the result of an unusual feature of the vegetative cell or of a morphological adaptation. Thus the pasteurization of milk can select vegetative cells having an above-average resistance to thermal denaturation, for example *Microbacterium lacticum*, *Streptococcus faecalis* and *Micrococcus* spp. Of course endospore-forming bacteria will survive also. Indeed when poor standards obtain in the cleaning of churns and storage tanks, milk can become heavily contaminated with *Bacillus cereus*. This organism's endospores survive pasteurization and, if storage conditions allow germination and outgrowth, outbreaks of 'bitty cream'

Table 6.5. Some exceptional implicit properties of microorganisms

Organism	Found in	Implicit properties
Lactobacillus trichodes	Wines containing 20% alcohol	Grows vigorously in wine 20% (v/v) ethanol; optimum initial pH range, 4.5–5.5
Lactobacillus fructivorans	Vinegar preserves	Maximum levels of the following growth-limiting factors given in parenthesis: acetic acid (4.2%); salt (12%), ethanol (18%); minimum pH 2.9
Streptococcus thermophilus	Cheese (a starter culture)	Optimum growth temperature, 40–45 °C, will grow at 50 but not 53 °C; survives 65 °C for 30 min
*Leuconostoc oenos**	Wine	Best growth in medium at pH 4.2–4.8; grows slowly in 18% ethanol at pH 4.8
Pediococcus halophilus	Soy mash	Best growth in 6–8% NaCl; tolerates 15% NaCl
Saccharomyces bailii	Acid beverages and fruit juices	Growth in acid medium containing 500 p/10^6 SO_2

* This organism is probably a species of another genus.

occur—'rafts' of cream float to the surface of hot coffee or tea to which contaminated milk has been added. These 'rafts' are formed because lecithinase produced by the vegetative cells modify the boundary of the milk fat micelle.

In appertization, the time/temperature relationship is such that only organisms with particular morphological adaptations survive; for example, the ascospores of yeasts such as *Saccharomyces bailii* and *Kluyveromyces bulgaricus* can survive the heat process given to acid fruit products. The ascospores of *Byssochlamys fulva* and *Bys. nivea* can survive the processing of canned strawberries—the fruits in infected cans are reduced to a purée during storage by pectinase produced by the vegetative cells that arise from the ascospores. With extreme thermal processing, the selective pressures favour the survival of endospores having exceptional heat-resistance, viz. *B. stearothermophilus*. The storage behaviour of a canned product will be determined by storage temperature. Thus endospores of *B. stearothermophilus* and *Desulfotomaculum nigrificans* will not germinate unless the storage temperature exceeds 40 °C (Fig. 4.3).

It may well be asked what value there is in identifying spoilage organisms or spoilage associations. There are at least two benefits to be gained.

1 Information is acquired about the spoilage of particular foods and thus an opportunity arises for devising specific media and methods to study

the occurrence and dispersal of spoilage organisms in the factory and by ingredients, as well as for monitoring their behaviour in a commodity both during preparation and storage.

2   It enables the definition of the constellation of implicit properties that fits an organism to a niche.

Not only will (1) offer the opportunities noted above but it will provide also a body of information that permits prediction of the most likely cause of spoilage of particular foods. In practice such information will aid the food technologist in the day-to-day management of food-processing operations and it may be expected also to provide an early warning should something unusual have occurred in the preparation of an ingredient or the food itself. With (2), a thorough understanding of an organism's properties may offer the opportunity for a rational approach to the choice of methods and processes for preserving particular foods.

Another facet of the spoilage story is the study of the chemical or physical changes which the microorganisms cause during the spoilage of a food. Again this may appear to be an academic exercise only. As with the characterization and identification of spoilage organisms, there are two possible benefits to be gained from a study of spoiled foods and the events leading to this state. A study of the events leading to spoilage may well lead in the future, when sophisticated systems of analysis are readily available, to methods which detect some early changes associated with the outgrowth of the spoilage organisms and thus a means of predicting quickly the possible interval before overt symptoms of spoilage are manifest. Such an analysis may also provide information about the nutrients used and modifications of intrinsic properties brought about by the spoilage organisms.

A study of spoiled foods will not only permit a definition of the chemical and physical changes associated with spoilage but it will enable the microbiologist to recognize those features which cause the would-be-consumer to reject a food.

It is possible with many spoiled foods to identify a major change which has been brought about by the spoilage organisms. Such changes may be one or more of the following:

(a)   *polymer* formation by the spoilage organisms—the polymers may be 'organized' as cells or occur as extracellular slime;

(b)   *depolymerization* by the spoilage organisms—carbohydrates (starch and pectin), proteins and fats may be the substrate and through their hydrolysis the biological organization of the food may be lost;

(c)   *destruction* of an emulsifier;

(d)   *fermentation* of a carbohydrate leading to the accumulation of fermentation products and scouring or merely a subtle but undesirable change in flavour;

(e)   *oxidation* of a product leading to a change in flavour or appearance;

(f)   *reduction* of a chemical resulting in an off-flavour or a change in colour;

(g)   *production* of a pigment;

(h)   *production* of an odour or off-flavour.

136

Examples of spoilage due to processes (a)–(g) are given in Table 6.6.

Little progress was possible with the chemical identification of off-odours until sophisticated systems of analysis were available. Indeed the older literature offers descriptions that can confuse rather than help the inexperienced microbiologist. What use is there, for example, in a spoilage odour being referred to as 'boiled cabbage' if this dish is not a part of the cuisine of the reader? Likewise there are perhaps several opinions on what

Table 6.6. Chemical changes—their spoilage symptoms and causative organisms.

Chemical change	Symptoms	Causative organism
(a) *Polymerization*		
Biological organization in cells	Slime on meat due to microbial growth at chill temperatures	*Pseudomonas fragi*
	Turbidity (haze) in wines, beers, ciders and beverages	Lactic acid bacteria and yeasts
	Yeast growth ('chalk moulds') on surface of bread	e.g. *Hypopychia burtonii*
Extracellular slime	'Ropey' milk	'*Alcaligenes viscosum*'
	'Ropey' cider	*Pediococcus cerevisiae*
	'Ropey' bread	*Bacillus subtilis*
	'Ropey' sugar products	*Leuconostoc mesenteroides*
(b) *Depolymerization*		
Destruction of biological organization	Maceration of canned strawberries	*Byssochlamys fulva*
	Maceration of carrots	*Rhizopus stolonifer*
(c) *Destruction* of an emulsifier	Bitty cream in pasteurized milk	*Bacillus cereus*
e.g. the breakdown of lecithin by lecithinase	'Custard yolk' in egg products	
(d) *Fermentation* of carbohydrate		
Accumulation of general fermentation products	Souring of milk	*Streptococcus cremoris*
	Souring of sausages	*Lactobacillus* spp.
Accumulation of a particular	Cider sickness (Acetaldehyde)	*Zymomonas mobilis* subsp. *pomaceae*
fermentation product (given in parenthesis)	Off-flavour in beer (Diacetyl)	*Pediococcus cerevisiae*
	Holes in hard cheese (H_2 CO_2)	Coliform organisms
(e) *Oxidation* of a product		
e.g. ethanol	Acetification of wines and beers	*Acetobacter* and *Gluconbacter* spp.
nitrosylmyoglobin	Greening of cured meat	*Lactobacillus viridescens*

(continued)

Table 6.6—*Continued*

Chemical change	Symptoms	Causative organism
(f) *Reduction* of a chemical		
NO_3^-	Blown cans of hams resulting from NO_3^- being used as a terminal electron acceptor	*Bacillus* spp.
SO_4^{2-}	Blackening (formation of ferric sulphide) of pickled vegetables	*Desulfovibrio* sp.
(g) *Production* of a pigment		
A phenotypic property of the organism	Red spot in cheese	*Lactobacillus plantarum*
	Pink spot on salted fish	*Halobacterium*
	Fluorescent-green whites in hens' eggs	*Pseudomonas putida*
	Purple spots in bread	*Bacillus subtilis*

constitutes a 'dirty dish cloth' smell and a certain botanical knowledge is presumably required in order to identify a 'may-apple' odour. As will be seen from Table 6.7, with the advent of g.l.c. and other methods of analysis, food odours can now be identified with specific chemicals.

Table 6.7. Chemicals that taint or flavour food.

Compound	Contributes to
Methane thiol $\{CH_3SH\}$	Off-odours of bacon, ham and fish Smell of some cheeses, e.g. the 'Cheddar aroma'
Trimethylamine $\{(CH_3)_3NH\}$	Through using the odourless trimethylamine oxide $\{(CH_3)_2N=O\}$ as an electron acceptor in the absence of O_2, some microorganisms produce trimethylamine which has a fish odour
3-methylbutanal $\{(H_3C)_2CHCH_2CHO\}$	Malty 'off' flavour in cheddar cheese. Some strains of *Str. lactis* contain transaminases and decarboxylases; these convert amino acids to aldehydes. Leucine is converted to the compound shown on the left.

Although many of the terms used to describe spoiled foods may as yet defy sensible definitions, the odours may well be the major reason for a would-be consumer rejecting a product, and the reason for rejection may be based on a *general* association within a community of people of a particular *odour* and spoilage. This is a feature which has to be considered by those who attempt novel methods of preservation of foods; such methods should have a 'fail-safe' attribute so that if spoilage has occurred it will be *recognized* by the would-be consumer.

7 Water

Water has many uses in the food industry. It may be used (1) as an ingredient; (2) to wash (and thereby decontaminate) ingredients; (3) to take heat to or away from a food, or (4) to clean equipment. For all these purposes the water must not contain pathogenic organisms (Table 7.1) and its content of food-spoilage organisms should be small. Much waste material, including human and possibly animal excrement, will be carried from a food factory in water. As heavy demands are made on the water resources of developed countries, much of the potable water needed by industry and communities will be reclaimed from sources that are persistently or occasionally contaminated with faecal material. In order that recycling of water can occur without causing epidemics among a population, there is a need for a sound body of knowledge about: (a) the role of water in the dissemination of pathogens; (b) effective methods of treating water that may contain faecal contaminants; (c) methods for monitoring the effectiveness of treatment systems, and (d) systems that can be used to treat sewage in order that its content of faecal microorganisms is diminished and its biochemical oxygen demand minimized so that effluent can be added to a stream, river or estuary without perceptible alteration of the ecosystems of the receiving water.

EPIDEMIOLOGY

Public attention was alerted to the possibility that water could 'cause' epidemics by John Snow's observations during the Golden Square (London) cholera outbreak in which 500 persons died within a 10 day period in 1854. He noted that the majority of victims resided within 250 yards of the public water supply, the Broad Street pump, but that none of the 70 workers in the Broad Street brewery, which had its own water supply, contracted the disease. It was noted also that a visitor from Brighton died following a 20 minute visit to his dead brother's house; the visitor drank a brandy and water. A woman who lived in another part of London employed a carter to collect daily water from the Broad Street pump; she died of cholera, as did a visiting niece. These snippets of evidence convinced Snow that the pump water was involved in the epidemic; indeed his conviction was such that he persuaded the vestrymen of St. James Parish to take the handle from the

Table 7.1. Water-borne pathogenic organisms.

Organism	Mode of transmission*	
	Direct	Indirect
Viruses		
Enteroviruses	+	+ ⎱ Faecally contaminated
Hepatitis A virus	+	+ ⎰ shell fish
Bacteria		
Salmonella typhi	+	+ ⎱ Faecally contaminated
Vibrio cholerae	+	+ ⎰ plant crops
Protozoa		
Entamoeba histolytica	+	+ ⎫
Giardia lamblia	+	+ ⎬ Faecally contaminated plant crops
Dientamoeba frugilis	+	+ ⎭
Nematodes		
Gnathostoma spinizerum		+ Present in raw or fermented freshwater fish
Angiostrongylus cantoniensis		+ Uncooked snails
Cestodes		
Diphyllobothrium latum		+ Uncooked freshwater fish
Sparganum spp.	+	+ Uncooked fish, frogs, *etc.*
Trematodes		
Clonorchis sinensis		+ ⎫
Opisthorchis spp.		+ ⎪
Metagonimus yokagawai		+ ⎬ Uncooked fish
Heterophyes heterophyes		+ ⎭
Paragonimus westermani		+ Uncooked crabs or crayfish
Fasciola hepatica		+ Faecally (sheep) contaminated watercress
Fasciola buski		+ Faecal contamination of water bamboo, water chestnut, lotus plant roots

* Direct, contamination via drinking water.
  Indirect, contamination of a food product by agencies listed above.

pump. This brought the epidemic to an abrupt end. The same strategy was adopted in 1959 to curtail an outbreak of typhoid fever among visitors to a camp site in Northern Ireland!

Direct proof of the potential of water to disseminate pathogens was lacking until bacteriology emerged as a discipline and indeed it was well into the present century before epidemiologists had finger-printing techniques (phage typing, for example—see Table 9.12) that permitted investigation of sporadic outbreaks in which a few people were infected. The majority of such outbreaks in the UK are due to *Salmonella typhi* (typhoid fever). Viruses, *Salmonella paratyphi* (enteric fever) and *Shigella* spp. (dysentery), are only occasionally responsible. The annual notification of typhoid fever in the UK (excluding Scotland) since 1936 is shown in Fig. 7.1. A marked decline in the notification rates between 1936–1945 was followed by a low but persistent level of outbreaks. In recent years the graph has shown a slight upturn due mainly to persons manifesting the disease in the UK after becoming infected in another part of the world, often the Indian subcontinent. In practice, of course, this contemporary trend provides interesting evidence of the potential of cheap and fast air travel to change slightly the epidemiological picture within a country (Fig. 9.3). This evidence is of particular interest to food microbiologists who are responsible in part for advising senior management on the policy to be adopted to safeguard products from contamination by staff who have visited areas in which typhoid fever is endemic.

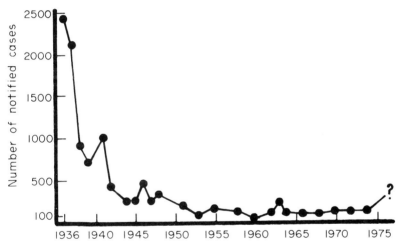

Fig. 7.1. The outbreaks of *Salmonella typhi* in the UK declined progressively in the first half of the century. A contemporary but modest upsurge in the trend has been attributed to outbreaks occurring among persons who contracted the disease in other continents. (Redrawn from *J. Hyg., Camb.* **81**, 139–49.)

Brief accounts of some of the outbreaks contributing to Fig. 7.1 are contained in the following section. In addition, a few pertinent facts are given for outbreaks of enteric fever and bacillary dysentery. These examples highlight the divers routes by which a pathogen can be transmitted from a depot of infection to susceptible hosts and draw attention to the contribution which faults in a water distribution system can make to outbreaks of gastroenteritis.

Typhoid fever

A carrier of *Salmonella typhi* resided occasionally at a house with an inadequate sewerage system which drained into a stream used by dairy cows. It was assumed that contaminants on the udders infected milk which was not pasteurized before distribution in the two towns. There were 518 cases in these towns and upwards of 200 among visitors who had returned home from their holidays during the incubation period.

CROYDON, 1937

It has been assumed that a chronic typhoid carrier was employed in the building work associated with a modification of the water distribution system; after modification, unchlorinated water was distributed and there were 310 cases in the town.

CORNWALL, 1941–1944

A man who had suffered from typhoid fever during the South African war was an occasional visitor to a hotel that had an inadequate sewerage system, the effluent passed near to a well from which water was pumped to storage tanks in the roof of the hotel. Several fatal cases of typhoid fever were traced back to the hotel.

EDINBURGH, 1963–1970

Occasional outbreaks among children playing in a stream were traced back via land drains connected to the sewerage pipe from the residence of a chronic carrier.

WEST MERSEA, 1950–1958

Untreated excrement from a carrier drained into an estuary from which shellfish were harvested.

ABERDEEN, 1964

Corned beef was contaminated with *Salmonella typhi* present in river water used to cool the cans post retorting.

Enteric fever

BRIXWORTH, 1941

Residents of condemned cottages obtained water from a well that was polluted by defective drains.

A chronic carrier contaminated sewage that found its way into a stream; infection passed to cattle and farm workers. A fractured pipe from the septic tank of the residence of an infected farm worker polluted the chlorinated water supplied to several villages in which many persons contracted the disease.

Dysentery and viral gastroenteritis

LEICESTER, 1952

Outbreaks in the workforce of a factory were attributed to the cross-connection of a pipe carrying potable water (at 25–30 lb pressure) and another carrying river water (at 240 lb pressure).

MONTROSE, 1966

It was estimated that 4000 of the town's population of 10 800 were affected. Town water was taken from a river polluted with sewage effluent from several villages, an aerodrome and a hospital along its course. After chemical precipitation of organic matter and positive-pressure filtration, the water was chlorinated (0.4 $p/10^6$) and stored for 3–10 h before distribution. Immediately before the outbreak, the chlorine level was 0.1 $p/10^6$ and *Excherichia coli* was present in water drawn from taps. Subsequent investigation identified a fault in the automatic switching gear which caused chlorine to pass from a full to an empty cylinder, thereby depleting the amount going to the water.

WATER TREATMENT AND DISTRIBUTION

Water of potable quality may be obtained from deep wells providing the shafts are impervious to seepage and the tops prevent surface water from draining into the well (Fig. 7.2). In this example, the water is freed of microorganisms as it filters down through the strata underlying the soil layer. With other sources, storage in reservoirs followed by chemical precipitation (e.g. aluminium sulphate) of organic matter, filtration and chlorination (with an appropriate holding period) provides water of potable quality. In the storage stage, many of the microbial contaminants die due to nutrient deprivation; many settle out; others are killed by phage or UV irradiation or engulfed by protozoa. Filtration plays an important role in breaking the infection route of protozoans, such as *Giardia* (Table 7.1), the cysts of which may retain their viability in chlorinated water under conditions that ensure the death of faecal bacteria.

Although storage, filtration and chlorination can produce potable water for a community or industry, these systems must be linked to an efficient

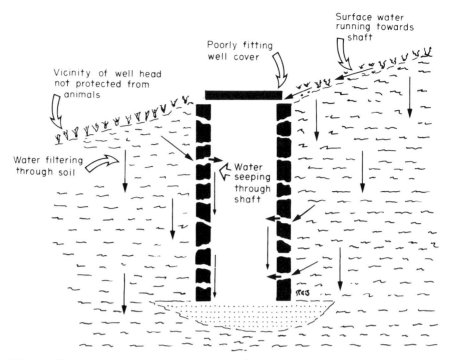

Surface water
running towards
shaft

Poorly fitting
well cover

Vicinity of well head
not protected from
animals

Water filtering
through soil

Water
seeping
through
shaft

Fig. 7.2. The water in a poorly sited and constructed well can be easily contaminated with microorganisms of human origin.

distribution network and the efficacy of the treatment and distribution systems must be monitored regularly. Even with a properly planned and constructed distribution system, microbial action may lead to deterioration of water quality. Thus microbial growth on the inner surfaces of pipes can cause the slow accumulation of sediments in which microorganisms grow because they are protected from chlorine. If relatively high temperatures obtain, such growth can initiate corrosion of the pipes or cause taints or off-odours in the water. Actinomycetes have been associated with the latter two faults. Moreover the growth of psychrotrophic *Pseudomonas* and *Flavobacterium* spp. in sediments could lead to the contamination of proteinaceous foods the preservation of which was based on chill storage. Problems due to sediments can be controlled by periodic flushing of the distribution system.

Materials used in a water distribution system can adversely influence water quality. Indeed it is now more than 50 years ago that a report appeared in which the unsatisfactory quality of a water supply was traced to the growth of coliforms on leather washers. As a diverse range of materials can be used for washers, 'O' rings and packing joints and glands, there is a need to assess a material's potential to support microbial growth before using it for one of these purposes. A recent survey of jointing materials showed that those based on vegetable fibres or oils or animal fats supported growth whereas polytetrafluoro-ethylene or silicone did not.

It is not uncommon for water to be given additional chlorine in a food factory. In-plant chlorination can control psychrotrophic microorganisms present in the water entering or during use in a plant. It is claimed also that this practice contributes to the maintenance of hygiene during working hours and limits the build-up of slime in the drains within a factory. When water is used for cooling, as in the chilling of broilers, quite high levels of chlorine (e.g. 50–200 p/10^6) may be added (Fig. 3.3). Likewise (see p. 86) chlorine is added to the water used to cool cans post retorting.

TESTING WATER

A critical assessment of a water supply calls for topographical, chemical, biological and microbiological investigations over a period of time so that seasonal variations can be monitored and the cause of related effects identified. In this section attention is given to the microbiological investigations only. With well and spring water in rural areas, microbiological tests provide a small community with some safeguards against water-borne pathogens. In general, however, the majority of investigations are concerned with quality control of water treatment and distribution systems. Although the actual tests vary in detail from one country to another, they have a common objective: to assay the level of contamination of water with organisms of or related to those of faecal origin. For convenience the methods used in the UK will be discussed in this section.

In order for the quality control of water to be effective, it is imperative that a supply is tested frequently by simple means rather than occasionally with a battery of complex tests. Moreover, as the number of pathogens in water is likely to be small, and thus the probability of their isolation is low unless large samples of water are examined in great detail, routine control is based on a search for intestinal organisms that are easily isolated and identified. Of the consortium of organisms in the gut of man and other mammals, coliform bacteria, faecal streptococci and *Clostridium perfringens* are of particular interest to the water microbiologist and emphasis is placed on the first-mentioned because they are the most sensitive indicators of water pollution—one such organism can be detected in 100 ml of water.

Coliforms are defined as Gram-negative, oxidase-negative, non-sporing rods which are able to ferment lactose with acid and gas production within 48 h at 37 °C. They grow aerobically on media containing surfactants. *Escherichia coli*, a member of this group, ferments lactose (and mannitol), with acid and gas production (Fig. 7.3), within 48 h at 44 °C; produces indole from tryptophane (Kovacs test, positive); does not have a citrate permease (no growth on a chemically defined medium with citrate as sole source of carbon); and carries out a mixed acid (methyl red test, positive; Fig. 7.4) but not the 2,3-butanediol fermentation (Voges-Proskauer test, negative; Fig. 7.5). These definitions contain those cardinal properties that have guided water microbiologists in their search for simple, rapid methods for the isolation and identification of these organisms.

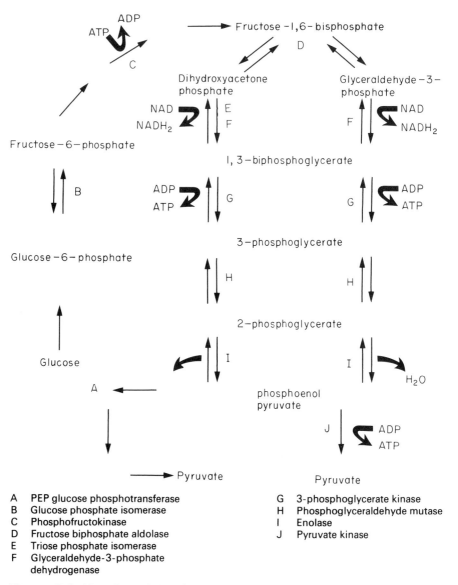

A	PEP glucose phosphotransferase
B	Glucose phosphate isomerase
C	Phosphofructokinase
D	Fructose biphosphate aldolase
E	Triose phosphate isomerase
F	Glyceraldehyde-3-phosphate dehydrogenase
G	3-phosphoglycerate kinase
H	Phosphoglyceraldehyde mutase
I	Enolase
J	Pyruvate kinase

Fig. 7.3. *Escherichia coli* uses the Embden-Meyerhof-Parnas pathway for the conversion of glucose-6-phosphate into pyruvate.

Historically the search for coliforms was based on the multiple-tube system and an elective/differential medium with incubation at 37 °C (Fig. 7.6). The assumption is made that each tube showing acid and gas after incubation at 37 °C for 48 h had been inoculated with one or more organisms in the measured volume of water added to the tube. Providing some of the tubes in the series shown in Fig. 7.6 do not exhibit these changes, the number of positive tubes can be used to determine (from probability tables) the most probable number (MPN) of the organisms being sought in 100 ml of water. The underlying theory is given below.

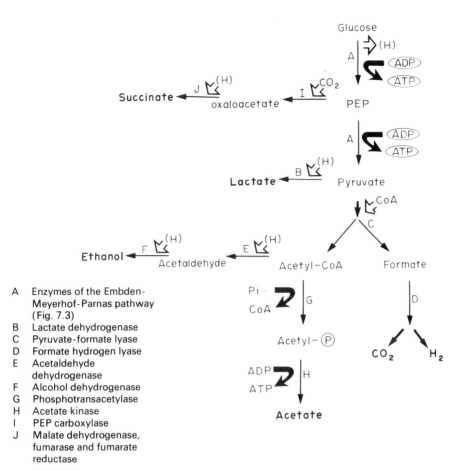

A   Enzymes of the Embden-
     Meyerhof-Parnas pathway
     (Fig. 7.3)
B   Lactate dehydrogenase
C   Pyruvate-formate lyase
D   Formate hydrogen lyase
E   Acetaldehyde
     dehydrogenase
F   Alcohol dehydrogenase
G   Phosphotransacetylase
H   Acetate kinase
I   PEP carboxylase
J   Malate dehydrogenase,
     fumarase and fumarate
     reductase

Fig. 7.4. The production of gas and a reaction that gives a positive methyl red test (pH 4.5) by *Escherichia coli* is due to the mixed acid fermentation outlined above.

Consider a sample of v ml taken from V ml of liquid containing x organisms. The probability of an organism being in the sample is v/V, of not being $(1 - v/V)$. From this the probability (P) that no organisms is in the sample is:

$$P = (1 - v/V)^x.$$

When v/V is small,

$$P = e^{-vx/V}.$$

As x/V is the density (ρ) of organisms ml^{-1},

$$P = e^{-v\rho}.$$

If n samples each of v are taken, and if s of these are found to be negative, the proportion (s/n) of samples is an estimate of P. Thus an estimate of the density (ρ) is given from

$$s/n = e^{-v\rho}$$

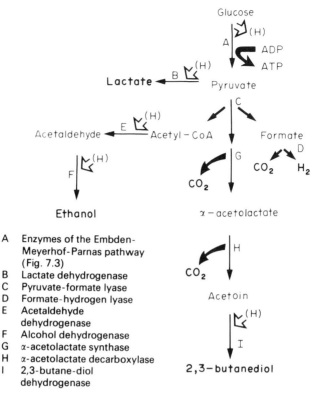

A   Enzymes of the Embden-
     Meyerhof-Parnas pathway
     (Fig. 7.3)
B   Lactate dehydrogenase
C   Pyruvate-formate lyase
D   Formate-hydrogen lyase
E   Acetaldehyde
     dehydrogenase
F   Alcohol dehydrogenase
G α-acetolactate synthase
H α-acetolactate decarboxylase
I   2,3-butane-diol
     dehydrogenase

Fig. 7.5. Unlike *Escherichia coli* (Fig. 7.4), *Enterobacter aerogenes* carries out a butanediol rather than a mixed acid fermentation.

which gives

$$d = (-1/v)\ln(s/n) = {}^-(2.303/v)\log_{10}(s/n)$$

where the estimate d is the most probable number of organisms per volume.

The commonly used series (1×50 ml; 5×10 ml and 5×1 ml of the water sample) is but one of a number of combinations that can be used to establish an MPN. As illustrated by Fig. 7.7, the precision of the method is improved by increasing the number of sub-samples per dilution examined.

The success of the MPN method is dependent also in large part upon the medium used to grow the organisms. It must be cheap but optimally elective for the organism in question; variability between batches must be minimal and the characteristic changes brought about by the organism must be clear-cut. MacConkeys Broth exemplifies the type of medium used for many years by water microbiologists for the isolation of coliforms. It contains peptone (an ingredient that is not easily standardized) as a nitrogen source, bile salts to select Gram-negative bacteria, and lactose as substrate for the fermentation leading to the production of acid (indicated by the pH indicator, neutral red) and gas (collected in a Durham tube). Selection of coliforms is favoured also by incubation (in a water bath) at 37 °C.

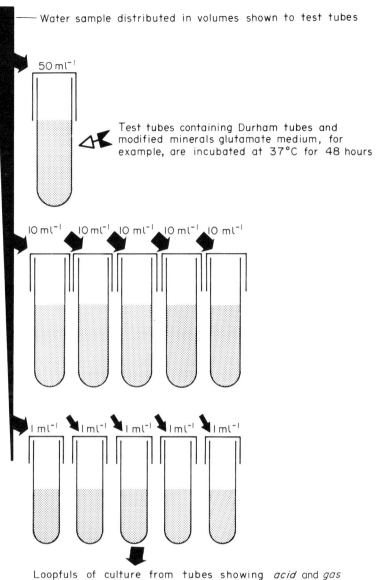

Fig. 7.6. The most-probable-number technique, with the combination shown above, can be used for the quantitative recovery of coliform organisms from water.

Difficulties in the standardization of bile salts led to the adoption of other surfactants, Teepol 610 in the UK and lauryl sulphate in the USA, for example. During the past 20 years a chemically defined medium (Minerals-Modified Glutamate Medium) has been adopted by many laboratories in the UK. In this case the enrichment of coliforms is the result of the careful selection of amino acids and vitamins in a mineral salts solution;

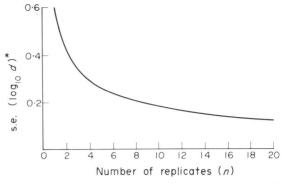

Fig. 7.7. The precision of the most-probable-number technique is influenced by the extent of replication, *viz*: the *Standard Error of the logarithm of the MPN density is inversely proportional to the square root of the number of tubes at a single dilution.

the diagnostic changes are still based on acid and gas production from lactose. When media such as those considered above are incubated at 37 °C, the most probable number of coliforms per 100 ml of water can be established because long experience has shown that in only rare cases is acid and gas production due to organisms other than those sought. Nevertheless the identity of the organism is presumed and in practice the reaction is referred to as a 'presumptive positive coliform reaction'. Several genera of the Enterobacteriaceae (e.g. *Escherichia, Citrobacter* and *Enterobacter*) are capable of producing acid and gas under the conditions of the test and in the UK further characterization needs to be done to establish whether or not *Escherichia coli* is present. With the rapid method of confirmation, each presumptive positive tube is subcultured to a tube containing lactose ricinoleate broth or brilliant green bile broth and incubated at 44 °C in a water bath for 24 h; the production of gas is taken to be a positive result. At the same time, subcultures are made to tryptone water; the presence of indole after 24 h at 44 °C is taken to be a positive reaction. Although recommended for routine use in confirmatory tests, it was recognized that brilliant green, an inhibitor of Gram-positive bacteria, was not easily standardized and ricinoleate tended to make a medium turbid thereby obscuring evidence of gas production. A recent survey indicated that these problems can be overcome by adopting lauryl sulphate, a surfactant that is chemically defined, as the inhibitory agent in a medium containing lactose and tryptophane so that only one tube, with incubation at 44 °C, needs to be used for confirmation of the presence of *Escherichia coli* in water. The sensitivity of the test could be improved marginally by substituting lactose with mannitol. It will be appreciated that for routine control purposes, an ideal situation has been achieved whereby confirmation of the presence of an indicator organism is obtained with a single medium.

A similar system to that used for the quantitative isolation of coliforms can be used, with appropriate media, for faecal streptococci (Fig. 7.8) and *Clostridium perfringens* (Fig. 7.9).

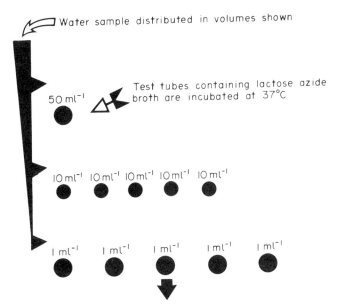

Fig. 7.8. The most-probable-number technique can be used for the quantitative recovery of faecal streptococci from water.

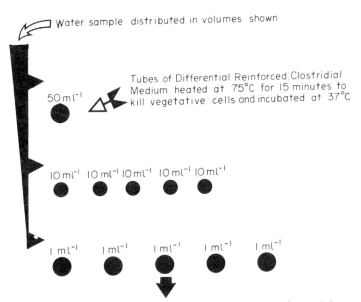

Fig. 7.9. The most-probable-number technique can be used for the quantitative recovery of *Clostridium perfringens* from water.

The introduction of membrane filters to microbiology allowed another approach to the counting of coliforms in water; the bacteria in a known volume of water remain on the surface of the membrane following filtration (Fig. 7.10). They form colonies when the membrane is incubated on a selective differential medium with an appropriate temperature regime. At the beginning of the routine use of this method in the UK, Teepol 610 was adopted as the selective agent but when commercial production of this compound ceased, alternative agents were sought. A recent trial indicated

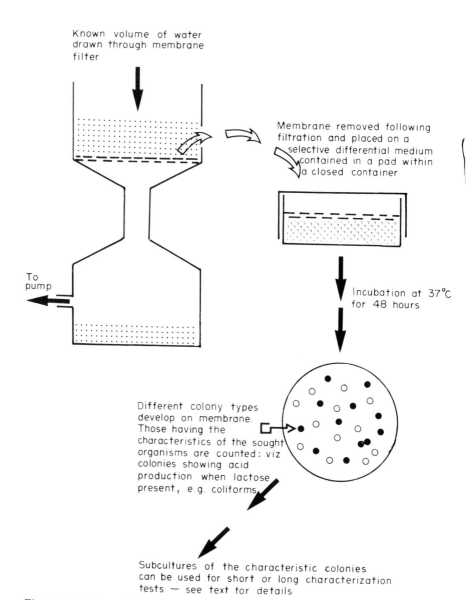

Known volume of water drawn through membrane filter

Membrane removed following filtration and placed on a selective differential medium contained in a pad within a closed container

To pump

Incubation at 37°C for 48 hours

Different colony types develop on membrane. Those having the characteristics of the sought organisms are counted: viz colonies showing acid production when lactose present, e.g. coliforms

Subcultures of the characteristic colonies can be used for short or long characterization tests — see text for details

Fig. 7.10. Membrane filters are used for the quantitative recovery of faecal microorganisms from water.

that there were no outstanding differences when Teepol was substituted by Tergitol T or sodium lauryl sulphate; the last-mentioned was recommended for routine use because it is cheap and chemically defined. The recommended medium contains peptone and yeast extract as a nitrogen/vitamin source and phenol red/lactose for differentiating the lactose-fermenting from other Gram-negative bacteria. The membrane filtration technique can also provide an opportunity for physiologically damaged or slow-growing organisms to form clones (Table 3.1), an initial 4-hour period of incubation at 30 °C is followed by 14 h at 35 or 37 °C or, for *Escherichia coli*, at 44 °C. After incubation the yellow (lactose positive) colonies are counted and further characterization attempted; for example, in the trial referred to above, material from a yellow colony on a membrane was subcultured into lactose peptone water (examined for acid and gas after 24 h at 37 °C) and on to MacConkey agar (purity, etc., checked) and nutrient agar (oxidase-positive organisms were presumed to be *Aeromonas* spp.). With membranes incubated at 44 °C, subcultures were made into lactose peptone water (positive result, acid and gas with incubation at 44 °C) and tryptone water (positive result, indole production at 44 °C).

In the above discussion, the identification of an organism with the coliform group or *Escherichia coli* was presumed. Should a particular investigation call for a substantive identification, then the tests set out in Table 7.2 would be used.

Table 7.2. The characterization of *Escherichia coli* and *Enterobacter aerogenes*.

	E. coli	Ent. aerogenes
Gas from lactose at 44 °C*	+	−
Mixed acid fermentation with glucose (methyl red, positive)**	+	−
Butanediol fermentation with glucose (Voges Proskauer, positive)†	−	+
Indole production from tryptophan	+	−
Growth on citrate as sole carbon source	−	+

* Fig. 7.3   ** Fig. 7.4   † Fig. 7.5
+ , positive result   − , negative result

In an ideal world, drinking water would not contain coliform organisms. As such perfection cannot be attained routinely, the significance of the occurrence of coliform organisms and *Escherichia coli* in water has to be assessed. Such an assessment must take into account the source of the water, its treatment and the population to be served. Thus in a rural area with a few households dependent upon a spring or a well, there would be grounds for concern if the coliform count exceeded routinely 10 per 100 ml of water and *Escherichia coli* was isolated repeatedly, especially if every effort had been taken to protect the water source from contamination. When unchlorinated water is distributed via pipes, then a small level of contamination of an occasional sample of water with coliforms (say, not more than 3 per 100 ml) can be accepted; the water would be deemed to be unsatisfactory if *Escherichia coli* was isolated. Coliform organisms must not be present in chlorinated water leaving a treatment plant. Such water can be contaminated during distribution and a tolerance has to be a part of any standard so that this situation is taken into account, viz. throughout any year coliforms and *Escherichia coli* must be absent from 95% of the samples. When the former do occur, they must not exceed 10 per 100 ml of water and no sample should contain more than 2 *Escherichia coli* per 100 ml. The frequency of sampling is of importance also and experience has shown that effective quality control is achieved with a scheme in which the size of a population supplied with water is inversely related to the intervals between sampling, viz. a monthly cycle for the water supplied to 20 000 people but daily for that piped to > 100 000.

If doubt is entertained about the results achieved in the routine examination of water for coliform organisms, then a search for faecal streptococci may prove helpful (Fig. 7.8). Thus with a water contaminated with coliforms but not *Escherichia coli*, the demonstration that it contains faecal streptococci would point to contamination with faecal material. As the endospores of *Clostridium perfringens* survive in water for longer periods than coliform organisms and as they tend to be resistant to chlorine at normal levels of use ($0.2-1.0 \text{ p}/10^6$ of free chlorine), their presence (Fig. 7.9) in natural water containing no coliforms suggests that faecal contamination of the source had occurred some considerable time before the sample was taken for examination.

Although the scheme outlined above will provide the food microbiologist with safeguards, his interests will extend also to organisms, *Pseudomonas* and *Flavobacterium* spp., for example, capable of growing under chill conditions. A plate count (yeast extract agar, 3 days at 20–22 °C) will give an indication of the content of such organisms in a water supply.

8 Sewage Treatment

It was noted at the beginning of Chapter 7 that the industries and communities of many countries are dependent on recycled water, the waste water of one community has to be converted into potable water for another. This calls for effective treatment of polluted water (Table 8.1) so that the following are avoided: (1) gross disturbance of the ecosystem of the receiving water and (2) illness in humans due to chemicals, waterborne pathogenic bacteria, viruses, protozoa and the eggs of round- and tapeworms (Table 7.1).

As the spent water of a community contains relatively small amounts of organic material (from a few to just a hundred or so parts per million), it would be theoretically possible to avoid the problem noted in (1) above simply by ensuring that gross dilution of the spent water occurred upon its addition to a stream or river. Many studies have catalogued the changes occurring downstream from a point receiving a constant outflow of sewage (Fig. 8.1). Of these changes, the receiving water's content of dissolved

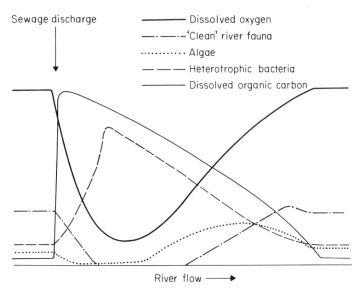

Fig. 8.1. Oxygen sag in a river receiving untreated sewage (graph from Lynch & Poole (1979) *Microbial Ecology : a Conceptual Approach*. Blackwell Scientific Publications, Oxford).

Table 8.1. Causes of pollution of natural waters.

Poisons

True poisons that tend to

    (i) accumulate in water and sediments, e.g. heavy metals—lead, arsenic, etc.

    (ii) have their toxicity enhanced by chemical change, e.g. mercury that is formed into organomercury compounds

Partial poisons e.g.

    (i) phenolic compounds that behave as antimicrobial agents when in large, but as substrates when present in minute amounts

    (ii) acids and alkalis that give a high but temporary concentration of hydrogen—or hydroxyl-ions—in a receiving water

Temperature

Elevation of the temperature of a receiving water causes two conflicting dangers, accentuated activity of the microflora and a diminished solubility of O_2 in water

Salts

Sodium chloride, for example, reduces the solubility of O_2 in water

Carbon wastes

Stimulation of microbial growth (eutrophication) and reduction of dissolved oxygen in a receiving water; the latter can progress to anoxia

Nitrogenous wastes

    (i) Contributory agents to eutrophication

    (ii) An end-product of nitrogenous waste breakdown, ammonia, exerts a high biochemical oxygen demand viz:

$$NH_4^+ + 2O_2 \rightarrow NO_3^- + 2H^+ + H_2O$$

    (iii) Ammonia a partial poison of fish viz: 3 mg m^{-3} is lethal to perch and rainbow trout over 2–20 h

    (iv) Ammonia reduces the efficacy of chlorine used as a disinfectant by causing chloramine formation viz:

$$NH_3 \rightarrow HOCl \rightarrow NH_2Cl + H_2O$$

    (v) Nitrates contribute to eutrophication and their presence in drinking water can also cause formation of methaemoglobinaemia in infants under six months of age—the 'blue baby syndrome'

---

oxygen is of particular interest. As Fig. 8.1 indicates, an oxygen sag is a typical feature downstream from an effluent pipe. It has been demonstrated that the well-being of fish, such as trout, is endangered if the oxygen concentration of water falls below 4 mg ml^{-1}, and this minimal value may be inadequate if the effluent contains toxic chemicals. Complete removal of oxygen would lead, of course, to a microbial flora dominated by facultative and obligate anaerobes, a collapse of the normal ecosystem of the river and the production of noxious smells. Providing the absorption of oxygen by a river is not impaired by either temperature, salts or both, the depletion of the oxygen dissolved in the receiving water will be correlated mainly with the load of organic material in the sewage. Thus in

Table 8.2. Analytical systems used in sewage treatment processes.

Biochemical oxygen demand (BOD)	Sewage sample diluted with aerated water and incubated in the dark at 20 °C for 5 d. The amount of dissolved oxygen at the beginning and end of incubation is determined by chemical methods viz. the Winkler test which is based on the reaction of oxygen with manganous oxide. An oxygen electrode can be used instead of a chemical method
Chemical oxygen demand	The oxygen absorbed by a sewage sample from boiling dichromate solution is determined. As this method exploits a strong chemical oxidation system, readily utilizable as well as recalcitrant (non biodegradable) organic molecules contribute to oxygen demand
Suspended solids	A sewage sample is drawn through a glass-fibre filter and the latter weighed

practice the 'strength' (Table 8.2) of a sewage and its likely impact upon a natural ecosystem will be determined by its biochemical oxygen demand (BOD). This can be measured with relative ease. The concentration of oxygen in a water sample is determined before and after five days' incubation (20 °C) in a blacked-out incubator—algal growth must be prevented—and the results expressed as the amount (mg) of oxygen consumed per litre of water. During incubation a microbial succession develops at the expense of the content of oxygen and organic matter in the water. The five day incubation period has been adopted in the UK because a Royal Commission on Sewage Disposal decided that this would be the maximum period required for water to pass along a British river to the open sea. Indeed it is the commission's standards that dictate the dilution rates that must obtain if sewage is 'treated' by dilution (Table 8.3). Although BOD is often criticized as an index for monitoring sewage, experience has shown that it is equal to if not better than many other indicators, viz. chemical oxygen demand.

Table 8.3. Royal Comission standards for sewage discharged into British rivers (from Lynch and Poole (1979) *Microbial Ecology: A Conceptual Approach*, Blackwell Scientific Publications, Oxford).

Available dilution by clean‡ river water	Standards (mg l^{-1})	
	BOD_5	Suspended solids
500	⋆	⋆
300–500	⋆	150
150–300	⋆	60
8–150	20	30
8	< 20†	< 30†

⋆ No standard recommended.

† Exact standard depends on local circumstances.

‡ Clean, usually means a BOD_5 of 2 or less.

If 'treatment' by dilution is not practicable, what alternatives are there? No simple answer can be given: thus the disposal of animal wastes on to fields matches the management requirements of farmers but much of England would be required should London adopt such a policy! In this century many systems have evolved to treat sewage: points common to most if not all such systems (Fig. 8.2) are that the BOD is diminished by:

(a) settling—suspended organic material is allowed to settle out through a reduction in the flow rate of the sewage;

(b) aeration.

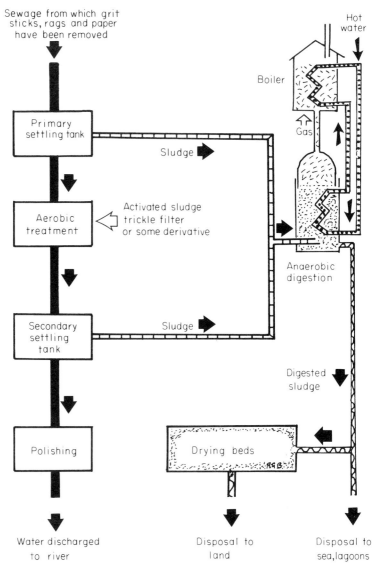

Fig. 8.2. A general flow diagram of a conventional sewage treatment plant. Each stage is discussed in the text.

The biodegradable solutes, colloids, etc. in the effluent water from the primary settling tank (Fig. 8.2) and a selected-adapted consortium of microorganisms are brought together under conditions such that the rate of microbial use of dissolved oxygen as a terminal electron acceptor never or only temporarily exceeds the rate at which it is replenished by mass transfer from a gas phase. The last-mentioned objective is achieved by (i) spraying fine droplets of the effluent into the atmosphere, or, more commonly, dispersing gas bubbles into the effluent. Activated sludge is the term applied to traditional methods based on (i) or (ii), or the two combined, when air and open tanks are used. Pressurized systems have led to improved efficiency, as with the 'deep shaft' method (Fig. 8.5), and the ready availability of 'high purity' oxygen, obtained by cryogenic or molecular sieve separation, has led to closed systems which are capable of dealing with very strong effluents from food processing units. The biological (trickle or percolating) filter is another method of bringing together the three elements in the aerobic treatment of waste-water, biodegradable materials, dissolved oxygen and microorganisms. The effluent from the primary settling tank, for example, is caused to form a thin film between the atmosphere and the colonized surface of some chemically inert material. As the organisms on the surface of the support medium are dosed periodically rather than continuously, the term 'intermittent feeding' will be used in the discussion of this method in this chapter because it avoids the confusion that can arise with the term 'filter' even when qualified as above. The most recent developments of treating sewage in fluidized beds represent a combination of essential elements of both the activated sludge and intermittent feeding systems.

As oxygen has such a cardinal role in the aerobic treatment of sewage, attention needs to be given to: (1) the methods that can be used to bring effluent into contact with this element; (2) factors that influence the transfer of oxygen across the interface between the liquid and the gas phase, and (3) factors that influence the movement of dissolved oxygen so that it is made available to microorganisms. As the solubility of oxygen in water is low, it is assumed that the transfer of oxygen from the interface into the under-lying liquid is rate limiting. The Oxygen Transfer Rate (OTR)—rate of oxygen transfer per unit time, per unit volume of liquid—is given by:

$$\text{OTR} = k_L \cdot a(C^\star - C_L)$$

where k_L = the liquid-phase mass-transfer coefficient; a (the specific surface) = the interfacial area per unit volume; $C^\star$ = the saturation constant given by the partial pressure of oxygen (P_B)/Henry's law constant (H_c), and C_L is the concentration of dissolved oxygen at the liquid gas interface.

It is evident from Fig. 8.2 that some of the organic material present in raw sewage is transformed into microbial cells, this biomass and the material harvested from the initial settling process tend to be glutinous and its disposal without further treatment difficult. Thus the anaerobic digestion of this material is an essential stage in most sewage works.

Vertical section through septic tank set in soil

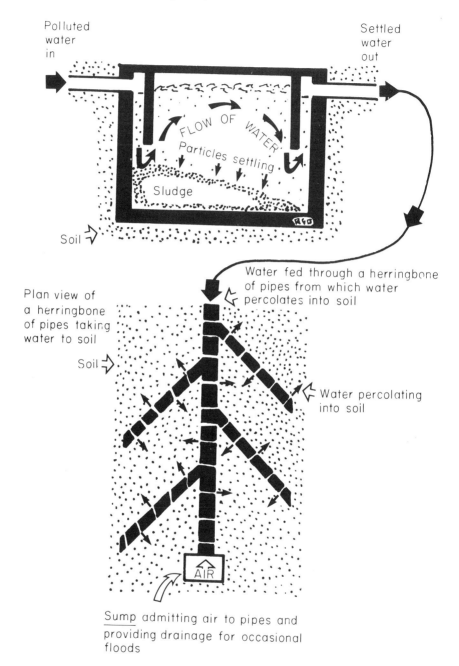

Polluted water in

Settled water out

FLOW OF WATER

Particles settling

Sludge

Soil

Water fed through a herringbone of pipes from which water percolates into soil

Plan view of a herringbone of pipes taking water to soil

Soil

Water percolating into soil

AIR

Sump admitting air to pipes and providing drainage for occasional floods

Fig. 8.3. Although a septic tank is a simple method for sewage treatment, it contains the elements exploited in large processing plants (Fig. 8.2) namely: (1) a reduction in flow rate causes suspended material to settle out as a sludge; (2) anaerobic digestion of the sludge with methane as one major end-product, and (3) aerobic treatment of the liquid phase by percolation through a porous soil.

These three main elements of sewage treatment are exploited in one of the simplest systems, the septic tank, which often provides the only practicable method for dealing with the waste coming from a few isolated houses in rural areas (Fig. 8.3). For the treatment of large amounts of sewage, the three elements noted above are separated one from another and the effectiveness of each enhanced by ancillary stages (Fig. 8.2); for example, the effectiveness of settling may be improved by chemical precipitation of the suspended materials and the possibility of gas-forming bacteria (e.g. *Clostridium* spp.) in the settled sludge impairing settling may be avoided by aeration of the raw sewage. Likewise, the collection of methane from the anaerobic digesters and its use in boilers to provide hot water to maintain the temperatures of digestion is another example of the efficiency of the principal process being bolstered by an ancillary one.

ACTIVATED SLUDGE

Effluent from the primary settling tank (Fig. 8.2) is passed along a tank fitted with diffusers (in the floor) or mechanical devices that cause violent agitation. By pumping air (or oxygen) through the diffusers or by agitation, oxygen dissolves in the effluent, the transfer rate from air at atmospheric pressure being of the order of 0.25 mmol O_2 l^{-1} h^{-1}. The dissolved oxygen supports the metabolism of obligate aerobes, such as *Zoogloea, Pseudomonas, Flavobacterium* and *Alcaligenes*, for example, that assimilate nutrients from the effluent; the vigorous agitation leads to the organisms forming flocs. Surface charges, possibly operating along with Ca^{2+} bridges, on polysaccharides enveloping the bacterial cells are the most likely mechanism of flocculation. In addition to the heterotrophs noted above, autotrophic microorganisms are included in the floc; for instance, nitrification of ammonia derived from the mineralization of proteins and amino acids is brought about by *Nitrosomonas* and *Nitrobacter*. Not only does the surface of a floc enmesh particles and absorb colloids, it provides a support for filamentous microorganisms, such as *Sphaerotilus natans*, and protozoans of which the ciliates appear to be of particular importance. Should a floc grow too large, the organisms at the centre die from oxygen deprivation; the cell autolysates are available to viable cells thereby accentuating the BOD of an effluent. When the effluent is no longer agitated and its flow rate diminished, the flocs settle out from the treated effluent (in the secondary settling tank) and a proportion of the flocs (activated sludge) is added to effluent from the primary settling tank at the beginning of the period of vigorous aeration. It will be appreciated that the activated sludge process can be considered as a continuous fermentation operated on a turbidostat principle.

It will be appreciated also that the activated sludge system (Fig. 8.4) can be regarded as an enrichment culture. In addition to the election of aerobic organisms by aeration of the effluent taken from the primary settling tank, the nutrient composition of an effluent will determine the phenotypic attributes of the organisms listed above and agitation followed by the

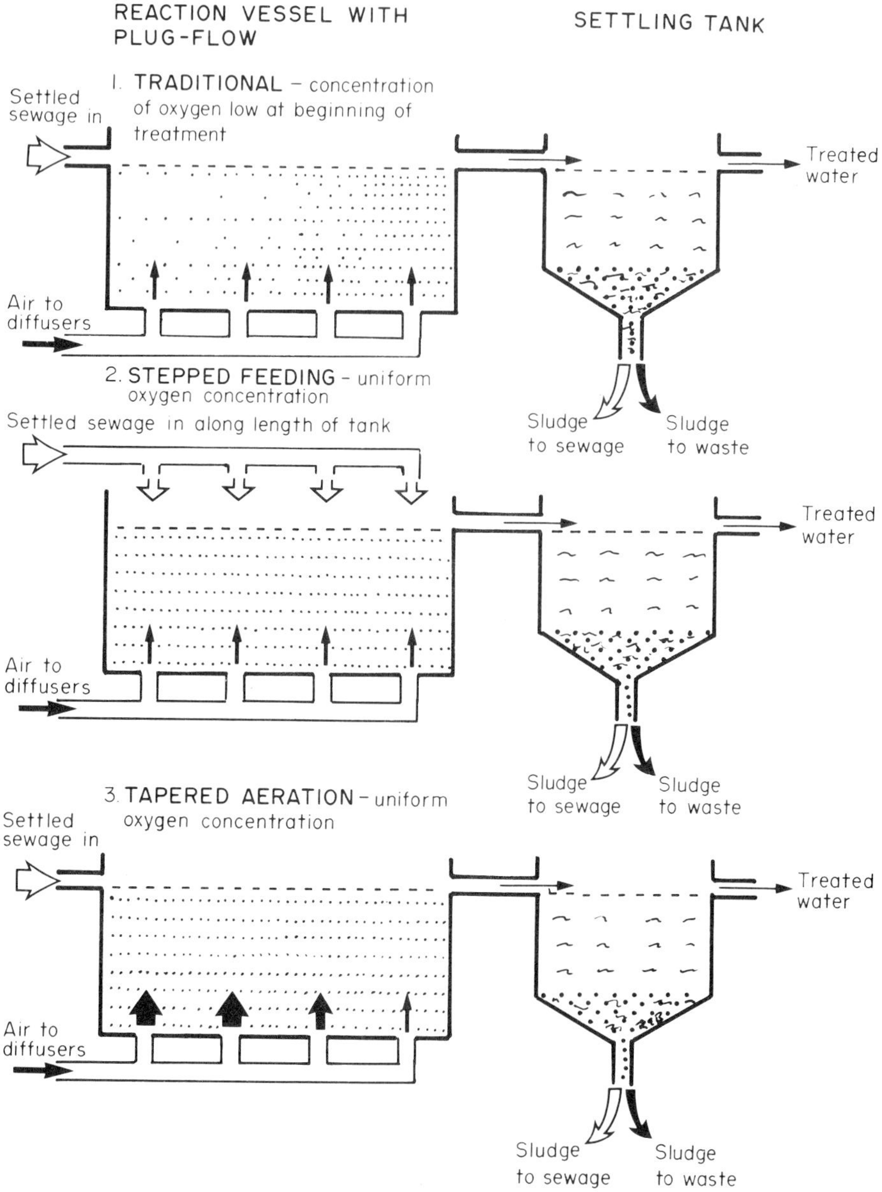

REACTION VESSEL WITH PLUG-FLOW

SETTLING TANK

1. **TRADITIONAL** – concentration of oxygen low at beginning of treatment

Settled sewage in

Air to diffusers

Treated water

Sludge to sewage   Sludge to waste

2. **STEPPED FEEDING** – uniform oxygen concentration

Settled sewage in along length of tank

Air to diffusers

Treated water

Sludge to sewage   Sludge to waste

3. **TAPERED AERATION** – uniform oxygen concentration

Settled sewage in

Air to diffusers

Treated water

Sludge to sewage   Sludge to waste

Fig. 8.4. Various systems can be used to ensure that adequate aeration occurs along the entire length of a vessel used in the activated sludge method of sewage treatment.

quiescent phase in a settling tank will select floc-forming rather than solitary organisms. Not only will the last-mentioned tend to remain in suspension and accompany the liquid phase to additional processing systems or discharge, but they will provide food for the predatory ciliates on the surface of the flocs. Indeed studies have shown that these protozoans play an important role in ridding settled sewage of its load of organisms of gut origin. It will be appreciated also that the activated sludge system will be vulnerable to toxic chemicals, especially if these are an occasional rather than a common feature of a sewage. In practice occasional 'shock loads' in a sewage can be due to the sudden release of phenolic compounds from a coking plant or metal ions, such as chromium, from a metal works.

Since the adoption of the activated sludge process in 1914, certain management systems and plant designs have been adopted to improve its overall efficiency. Although basically this system of sewage treatment is characterized by a long retention time of the floc-forming organisms and a short residence time of the liquid, two systems of management can be considered. As 90% of the BOD is removed extremely quickly from settled sewage by adhesion of particles and absorption of colloids to the floc surface and the diffusion of soluble material into the flocs, a relatively large inoculum of activated sludge can be added to effluent coming from the primary settling tank. After a short period of contact, the flocs can be allowed to settle out and are then aerated so that assimilated materials are broken down to a point when endogenous respiration is induced and the mass of the sludge reduced, thereby minimizing the amount that needs to be digested anaerobically (Fig. 8.8). Alternatively a smaller inoculum can be used with a long residence time (24–48 h) of the effluent taken from the primary settling tank (Fig. 8.2) so that the initial rapid removal of BOD, breakdown of absorbed material and endogenous respiration occurs in a single process. With either approach, some wastage of sludge is essential otherwise a progressive increase in the mass of a floc will cause necrosis at its centre and an increase in the overall BOD of the process. With plant design, attention has been given to (1) agitation so that contact between flocs and soluble and suspended materials in the effluent is maximal and (2) aeration. With the latter, two major strategies have been adopted in order to ensure that oxygen does not become a limiting factor. With the inappropriately named system, stepped aeration, effluent is fed in along the length of a rectangular aeration tank in an attempt to get a uniform oxygen demand. In the other approach, tapered aeration, the effluent is fed in at the end of a tank and the amount of air pumped into the liquid decreases along the length of the tank (Figure 8.4). Not only will a low concentration of dissolved oxygen impede the processing of sewage effluent, it can contribute to the fault known as 'bulking' sludge. Such sludge fails to settle in the secondary settling tank because the surface of the flocs are excessively colonized by filamentous microorganisms, such as *Sphaerotilus natans*, which, in addition to storing fats, act as suspensory organelles.

This fault can be caused also by temperatures exceeding 30 °C, an imbalance of the rates of nutrient absorption to the flocs (high) and microbial catabolism (low), and acid conditions promoting the growth of moulds. It will be appreciated also that the dense populations of microorganisms in settling sludge will remove dissolved oxygen quickly. With anoxic conditions, anaerobic respiration with NO_3^- as terminal electron acceptor will be favoured and bubbles of nitrogen and nitrous oxide formed (Fig. 8.6). If such bubbles become trapped in the flocs, the sludge will rise up in the settling tanks.

Increasing oxygen transfer rates

As the oxygen transfer rate (see p. 159) determines in large part the efficiency of activated sludge methods, any strategy that increases this rate will benefit a treatment system. Thus an increase in pressure above atmospheric on air will increase the partial pressure of oxygen and hence the driving force for mass transfer across the liquid–gas interface. This principle is exploited in the 'deep shaft' method (Fig. 8.5) in which a reduction of $>90\%$ BOD can be achieved with the retention of 'strong effluent' $(3.5–9.5$ kg BOD m^{-3} d$^{-1})$ for as little as 2 h. Preliminary studies indicate that the flora developing in the shaft is not dissimilar from that in conventional methods for activated sludge. The pressure in the shaft may well modify the morphology of *Sphaerotilus*, the organisms occurring in such short chains that they are unable to function as suspensory organelles.

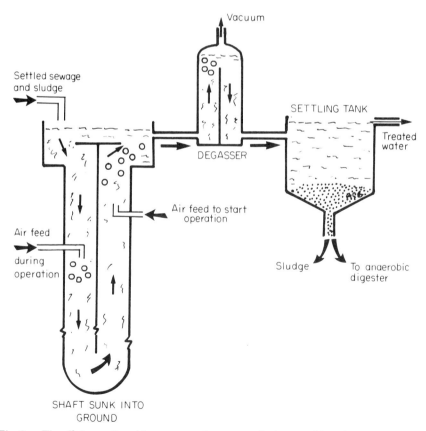

Fig. 8.5. The efficiency of aerobic treatment of sewage can be improved by using air under pressure as in the 'deep-shaft' method (see text for details).

Indeed the sludge coming from the shaft has excellent settling qualities if the flocs are freed of the small bubbles of gas which form when nitrogen, carbon dioxide and any unused oxygen come out of solution with the drop in pressure.

Another way of increasing the driving force for mass transfer of oxygen across the liquid–air interface is to use 'high purity' oxygen. Not only can 'high-purity' oxygen be used to increase the efficiency and hence the operating capacity of existing systems of activated sludge, it can be considered as a means of coping with seasonal peaks in effluent flow into plants that are working at or near maximum capacity for much of the year. In the final analysis, of course, the best results are likely to be achieved with a closed system that minimizes wastage of the 'high-purity' oxygen. In theory a combination of pressures greater than atmospheric and 'high-purity' oxygen might be expected to provide the most efficient systems for the aerobic treatment of waste-water. As yet no such system has been adopted commercially, mainly because of uncertainty about microorganisms' tolerance of high concentrations of dissolved oxygen.

INTERMITTENT FEEDING

This method can be said to be a derivative of one of the oldest systems used to dispose of the waste of a community, the intermittent application of the liquid waste of settled sewage to farm land—a practice that led to the term, sewage farm, being introduced to the English language. In this instance, a heterogeneous soil flora present in a medium of sand, humus and clay, in which the void spaces are of very small dimensions, assimilates and degrades the organic material in the effluent. With contemporary processes, feeding is still intermittent but the organisms form a film on the surface of an inert substrate (Fig. 8.7). Indeed the following discussion draws attention yet again to an industrial process which is founded, in large part, on the principles of elective culture. By trickling water containing nutrients over an inert surface exposed to the atmosphere, a consortium of micro- and macro-organisms is selected because of the ability of some of its members, the pioneers, to adhere to the inert surface and resist erosion. Other members exploit the pioneering organisms' ability to adhere to the surface whilst others live in or on the film by predation (protozoa) or grazing (fly larvae). Although numerous attempts have been made to identify members of the consortium, the literature tends to present a catalogue of organisms recovered, much comment about the many unsatisfactory features of microbial classification but little detailed information about the precise contribution of named organisms to the overall functioning of the film.

Many factors contribute to the maintenance and influence the attributes of a film: insolation; temperature; the volume and rate at which liquid is added (hydraulic loading); the nutrient content of the liquid; the distances between opposing films and the tortuosity, or otherwise, of the interconnecting void spaces between films; the thickness of the film and the availability of oxygen. With insolation, for example, algae and moss will grow and the treatment process will tend to be productive rather than degradative in nature. Not only does ambient temperature influence the metabolic rate of the film, it will influence thickness also. Indeed an inverse correlation of film thickness and temperature has been noted; during spring a relationship of the quantity of film to the amount of nutrient in an effluent of 0.5 g film g^{-1} of BOD was noted whereas in summer the value fell to 0.08 g g^{-1} BOD. This reflects not only the higher endogenous respiration rate of a film during the summer months but also the increased activity of the grazing members of the film, especially the fly larvae. Thus in practice an interplay of temperature and nutrient content of an effluent will dictate the distances between opposing films and, should film growth be excessive, the void spaces obtaining between them will be occluded. This can lead to a temporary perturbation of gas and liquid flow or the permanent blockage of water flow, a state referred to as 'ponding'. Not only the strength of an effluent, but its rate of application will influence the characteristics of a film. With high rates of application, the film is scoured and the void spaces freed of debris. Film thickness needs to be considered also in a context other than that of space between opposing films. As a film increases in thickness, the availability of oxygen to organisms at the interface of film and the surface of the support medium will become less and less. This can lead to the death of these organisms and the sloughing of the film. It is possible also for sloughing to be accentuated by facultatively anaerobic bacteria; when denied oxygen, such organisms become fermentative and the gas they produce forces the film away from its support. In the final analysis, however, the performance of a trickle filter at any one loading rate will be determined by the area available for colonization by the film-producing organisms and the void spaces available for the flow of liquid and gases.

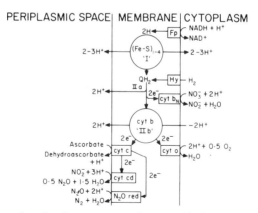

Fig. 8.6. Organisms such as *Pseudomonas* spp. respire anaerobically with nitrate as terminal electron acceptor under conditions of anoxia. This figure shows the electron flow in anaerobic respiration with NO_3^- (otherwise known as denitrification). Redrawn from Stouthammer A.H., Boogerd F.C. & van Versveld H.W. (1982) Bioenergetics of dinitrification. *Ant. van Leeuwenhoek*, **48**, 545–553.

In its traditional form, the trickle filter (Fig. 8.7) of a sewage plant consists of angular pieces of clinker, granite or slag of 50–100 mm size packed randomly to a depth of 2 m and fed effluent from the primary settling tank via holes or jets in the trailing edges of rotating arms. In such a filter there will be upwards of 80–110 m² of surface available for microbial colonization per m³ of the support medium and about 50% of this volume will be available for liquid and gas flow. The height of 2 m is dictated by mechanical considerations; any advantages to be gained from an increase in height would be offset by the increased costs of providing an adequate, load-bearing foundation. The number of filters and/or their diameters are the principal variables in the provision of trickle filters of the traditional type for the processing of effluent from a community. Thus management assumes an important role in the maintenance of film thickness and several options can be considered. With relatively weak effluent giving a loading of about 0.09 g BOD m^{-3} d^{-1}, it can simply be fed to the filter, collected in the secondary settling tank, and material that settles out is transferred to the anaerobic digester. With stronger effluent (about 0.2 kg BOD m^{-3} d^{-1}), that taken from the primary settling tank can be diluted by about 2:1 with liquid from the secondary settling tank. In cases of even stronger effluent, two applications of the effluent to trickle filters can be considered. In one approach the effluent from the primary settling tank can be applied to a filter containing coarse (75–110 mm) aggregate without fear of the void spaces becoming occluded. The effluent from this filter is then applied to a second filter in which smaller aggregate is used. Alternatively a 'feast-fast' regime can be operated. Strong effluent is applied to a trickle filter until the growth of film begins to impair water movement through the aggregate; the effluent is then applied to a second filter, in which the film has been starved by receiving effluent from the first filter. The latter will now be treated with filtered effluent and the film will decline as a consequence of endogenous respiration.

Many of the constraints imposed by the materials used in a traditional trickle filter have been overcome by adopting plastics to support the microbial film. It is used as sheets, the topography of which is designed to give maximum surface area per unit volume, which are separated one from another by a distance that ensures freedom from ponding and maximum gas flow. Alternatively the plastic is used to form small units of complex shape and these are used to pack filters. In whatever form it is used, the low density of plastic allows its use in tall, tower-shaped filters rather than the squat (2 m) ones of the traditional sewage plant. Moreover, tall, plastic-filled filters permit the treatment (loading rates 1.7–5.0 kg BOD m^{-3} d^{-1}) of the very strong effluents of the food industry.

In the past few years, attention has been given to another strategy, taking the film of organism to the effluent rather than feeding the effluent on to the immobilized film of a trickle filter. Discs, separated from one another by a short distance, are fitted to an axle and rotated slowly so that about 40–50% of the surface of each disc is submerged in effluent at any one

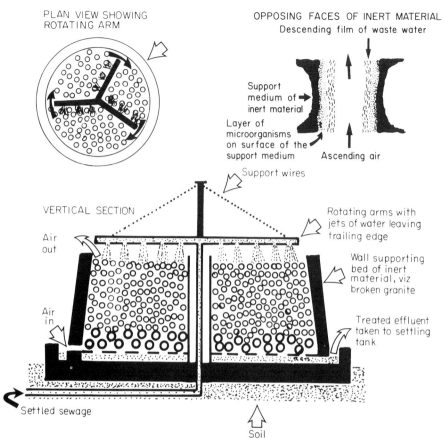

PLAN VIEW SHOWING
ROTATING ARM

OPPOSING FACES OF INERT MATERIAL
Descending film of waste water

Support
medium of
inert material

Layer of
microorganisms
on surface of the
support medium

Ascending air

Support wires

VERTICAL SECTION

Rotating arms with
jets of water leaving
trailing edge

Air
out

Wall supporting
bed of inert
material, viz
broken granite

Air
in

Treated effluent
taken to settling
tank

Settled sewage

Soil

Fig. 8.7. A traditional trickle filter.

time. Rotation causes complex flow patterns in the liquid between discs and a very efficient transfer of nutrients from the effluent to the micro-organisms colonizing the surface of the discs. It will be appreciated that this method results in the film at any one location on the disc being alternatively submerged in effluent, thus assimilating nutrients, and exposed to the atmosphere so that the assimilated material can be mineralized to provide energy for transformation of some of the material in the waste to cell material which, in turn, will be broken down by endogenous respiration. In a derivative of this system, a cylinder formed from a heavy metal mesh and filled with plastic units of complex geometry is rotated in effluent so that only a proportion of the plastic units are submerged at any one time. In addition, this system is so constructed that those plastic units that are not submerged have effluent trickled over their surfaces. This system has proved effective in treating strong effluents.

FLUIDIZED BEDS

The evolution of 'fluidized-bed' reactors has seen a marriage of certain of the features of the activated sludge systems and those of the intermittent feeding methods. Thus a solid phase of inert material, sand or plastics, are loosely packed in a column fed from beneath with waste-water and supplied with air or 'high-purity' oxygen. The upward flow of effluent is regulated so that sand or other material is kept in suspension. The solid phase becomes clothed with microorganisms but liquid flow causes attrition between the individual particles with the consequence that the bed is prevented from becoming blocked. Moreover the colonized particles break up gas bubbles introduced at the bottom of the bed. This results in a very favourable ratio of surface area of bubble to volume of liquid needing oxygenation.

The water collecting in the secondary settling tank after treatment by the activated sludge or biofiltration systems (Fig. 8.2) will probably be of a quality that allows its discharge to a river or stream. If such a quality has not been attained, some additional treatment ('polishing') will be given; for example, the water may be filtered through sand or pressure-filtered to remove particulate matter and chlorinated before discharge. When there is evidence of a disturbance of the ecosystem of the receiving water due to nitrates in the effluent, then the effluent can be dosed with a non-fermentable substrate, e.g. methanol, and stored without access to oxygen so that denitrifying microorganisms, such as *Pseudomonas* spp., respire anaerobically with NO_3^- as the terminal electron acceptor (Fig. 8.6).

Nitrification

$$NH_4^+ + 1\tfrac{1}{2}O_2 \rightarrow NO_2^- + 2H^+ + H_2O$$

$$NO_2^- + \tfrac{1}{2}O_2 \rightarrow NO_3^-$$

Denitrification

$$\text{Carbon substrate} + 2H_2O + 4NO_3^- \rightarrow 2N_2 + 4OH^- + 5CO_2$$

SLUDGE AND ANAEROBIC DIGESTION

The material (sludge) collecting on the floor of the primary and secondary settling tanks (Fig. 8.2) is malodorous and of such a glutinous nature that water loss by drainage and evaporation can take many months. Thus if dumping (e.g. at sea) is not practicable or lack of space precludes storage in lagoons for up to 2 years, the sludge has to be digested, drained and disposed on farm land, in waste tips or by incineration. It will be appreciated that sludge, a very concentrated slurry of organic materials when compared to the dilute solution in the water taken from the primary settling tank, will need a relatively long time to be digested by methods that are not dependent upon aeration. Indeed digestion by anaerobic bacteria in closed tanks is not completed in much less than 60 days at the temperatures obtaining in the northern and southern hemispheres. Facultatively anaerobic bacteria probably contribute little if anything to digestion of the sludge but they may scavenge O_2 entrained in fresh sludge and thereby cause a drop in the redox so that obligate anaerobes, such as *Clostridium* spp., proliferate.

The last-mentioned hydrolyse proteins, polysaccharides and fats and ferment the monomeric by-products and thereby provide the substrates, acetic acid, H_2 and CO_2, for methane production by the strictly anaerobic methanogenic bacteria (Fig. 8.8).

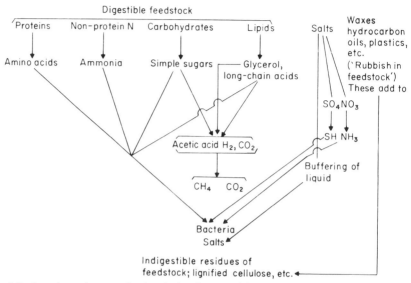

Fig. 8.8. Steps in methane production during the anaerobic digestion of sewage sludge. (From Lynch & Poole (1979) *Microbial Ecology: A Conceptual Approach*, Blackwell Scientific Publications, Oxford.)

By fuelling boilers with methane harvested from the anaerobic digestion of sludge, hot water can be circulated through the digestion tank and, by maintaining temperatures of 25–30 °C, the retention period reduced to 20–30 d (*cf.* 60 d at ambient temperatures). Although laboratory studies of the chemical attributes of the clostridia and methanogenic bacteria allow a succinct description of the anaerobic digestion of sludge, the day-to-day management of a digestion tank calls for methods that ensure a balance between the production of acids, especially acetic acid, by clostridia and methane by the methanogenic bacteria. This balance can be disturbed by adding sludge to the digester at too fast a rate, allowing the concentration of solids to exceed 10% or by failure to mix the contents of the digester thoroughly so that stratification results in a solids-rich sediment with an extensive surface of foam and a temperature gradient throughout the tank. As a rule of thumb, the acid concentration (measured as acetic acid) of the digesting sludge should not exceed 500 mg l^{-1}.

When digestion is completed, the water accumulating on the surface of unstirred sludge is pumped into the primary settling tank (Fig. 8.2) and the digested sludge transferred to a bed of clinkers on an inclined concrete platform. With storage, the sludge loses water by drainage and evaporation and, after 3–6 months, it is in a form (a brown, friable material) suitable for disposal. Although application to farm land would appear to be the most sensible policy, particularly as the digested sludge acts as a fertilizer as well as a soil conditioner, the risk of poisoning farm land with metal ions, e.g. copper and zinc, must be considered. There is also the possibility of disseminating pathogenic organisms.

9 Food Mediated Disease

If illness follows shortly after the consumption of food, then it is customary for the victim to claim that he or she has food poisoning. When such a general definition is adopted, a diverse range of causative agents of food poisoning can be listed (Table 9.1). This chapter deals mainly with only one major category in this list, bacterial food poisoning. This is characterized by illness (Table 9.2) which follows the consumption of food containing specific bacteria that have formed: (1) a population large enough to colonize the gut of a susceptible host; (2) a population large enough to intoxicate the gut of a susceptible host, or (3) clinically significant amounts of toxin. During the past 20 years or so, it has been demonstrated that many fungi can elaborate pharmacologically active substances (mycotoxins) which if they occurred in foods could pose a threat to the health of man, a topic discussed at the end of this chapter. In many instances, detailed clinical studies of a victim of suspected food poisoning, pathological examinations of his/her faeces, blood, etc., and laboratory investigations of the food(s) consumed before the onset of symptoms fails to identify the aetiological agent. In many cases viruses are the probable agents and the failure to identify them is due mainly to the lack of adequate methods of isolation.

BACTERIAL FOOD POISONING

In theory, Koch's postulates ought to be met before a bacterium is identified as a food-poisoning organism. In practice this situation obtains with *Salmonella* spp., *Vibrio parahaemolyticus*, *Escherichia coli*, *Yersinia enterocolitica*, *Clostridium perfringens* and *Campylobacter jejuni*. With these organisms, a causal relationship between a specific microorganism and a specific disease can be established thus: (a) the microorganism is present in every case of the disease; (b) the organism can be isolated from the victim of food poisoning and grown in pure culture; (c) the specific disease symptoms are reproduced when a pure culture is fed to human volunteers or a susceptible host, and (d) the organism can be re-isolated from the experimentally fed host.

Evidence of a more circumstantial and epidemiological nature may be used in studies of food poisoning caused by *Staphylococcus aureus*,

Table 9.1. The causative agents of food poisoning.

Illness follows immediately or shortly after the ingestion* of a food containing	Examples
A poison:	
added accidentally during the preparation of a food	Lead poisoning due to large amounts of element in food
Synthesized by a commodity during growth or post-harvest storage	Solanine poisoning caused by alkaloids in potato. Red bean poisoning caused by a haemagglutinin that is not destroyed by inadequate cooking
formed during the processing/storage of a food under poorly controlled conditions	Scombrotoxic fish poisoning caused by the production of a moderately resistant compound in fish stored under unsatisfactory conditions. It causes a 'histamine-like' reaction; caused by fish such as mackerel
stored by a commodity	Paralytic shellfish poisoning caused by a toxin of the dinoflagellate, *Gonyaulax tamarensis*, which is stored in mussels Ciguatera poisoning caused by warm-water, carnivorous fish which acquire the poison from herbivorous fish which in turn obtained it from plankton/algae
An allergen	An allergy to strawberries
Bacteria that grow in a food and form populations large enough to:	
colonize the gut of a susceptible individual	*Salmonella* spp. *Vibrio parahaemolyticus* *Escherichia coli* *Campylobacter jejuni* *Yersinia enterocolitica*
intoxicate the gut when eaten by a susceptible host	*Clostridium perfringens*
produce sufficient toxin† in the food to cause illness in a susceptible host	*Staphylococcus aureus* *Bacillus cereus* *Clostridium botulinum*
Unidentified causal agents	Gastroenteritis of unknown origin; viruses are commonly suspected to be involved

* By stressing that the incubation period is short and that the alimentary canal is the primary target of the physiological changes causing the illness or the site from which toxins are absorbed, an attempt has been made to separate diseases that occur in a susceptible person consuming contaminated food from those in which microbial growth and/or toxin production in a person are essential prerequisites for the disease syndrome.

† The available body of information tends to be concerned mainly with bacterial toxins because epidemiological studies have been able to establish a link between the toxins and the abrupt onset of well-characterized clinical symptoms. There is an equally large body of information on the toxins (mycotoxins) produced by fungi (Table 9.13) but at the moment the epidemiological evidence is not sufficiently detailed to permit the acceptance of numerous suppositions that certain diseases are due to these toxins (see p. 218).

Table 9.2. Clinical features of the illnesses produced by food-poisoning bacteria.

Bacteria	Illness
Infection types	
Salmonella spp.	INCUBATION PERIOD, 6–48 h; usually 12–36 h DURATION, 1–7 d Diarrhoea, abdominal pain and vomiting Fever nearly always present
Vibrio parahaemolyticus	INCUBATION PERIOD, 2–48 h; usually 12–18 h DURATION, 2–5 d Diarrhoea, profuse and often leading to dehydration; abdominal pain and fever
Escherichia coli	INCUBATION PERIOD, 12–72 h DURATION, 1–7 d (a) Cholera-like illness with watery diarrhoea and pain (b) Prolonged diarrhoea with stools containing blood and mucus
Campylobacter jejuni	INCUBATION PERIOD, 2–10 d DURATION, 5–7 + d Flu-like symptoms with abdominal pain and fever followed by diarrhoea, often severe
Yersinia enterocolitica	INCUBATION PERIOD, ? DURATION, upwards of 3 d Abdominal pain, fever, headache, diarrhoea, malaise, vomiting, nausea and chills. Acute lymphadenitis, resembling appendicitis, in children has led to appendectomies
Toxic types	
Clostridium perfringens	INCUBATION PERIOD, 8–12 h DURATION, 12–24 h Diarrhoea, abdominal pain, nausea but rarely vomiting; no fever
Staphylococcus aureus	INCUBATION PERIOD, 2–6 h DURATION, 6–24 h Nausea, vomiting, diarrhoea and abdominal pain; no fever. Collapse and dehydration in severe cases
Bacillus cereus	INCUBATION PERIOD, 8–16 h DURATION, 12–24 h Abdominal pain, *diarrhoea* and occasionally nausea INCUBATION PERIOD, 1–5 h DURATION, 6–24 h Nausea and *vomiting*, and occasionally diarrhoea
Bacillus licheniformis★	INCUBATION PERIOD, 4–8 h DURATION, ? Diarrhoea and vomiting
Bacillus subtilis★	INCUBATION PERIOD, 2–18 h DURATION, ? Vomiting, abdominal pain, diarrhoea
Clostridium botulinum	INCUBATION PERIOD, usually 18–36 h Death in 1–8 d or slow convalescence over 6–8 months. Symptoms variable: disturbance of vision, difficulties with speaking and swallowing. Mucous membranes of mouth, tongue and pharynx usually dry. Progressive weakness and respiratory failure

★ Organisms for which evidence is only beginning to accumulate.

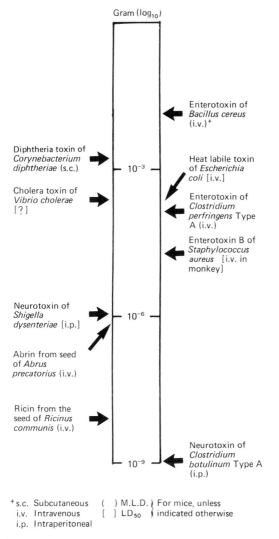

Fig. 9.1. The relative toxicity of bacterial and plant toxins (taken from Gill (1982) *Microbiological Reviews*, **46**, 86–94).

Clostridium botulinum, Bacillus cereus and other toxin-producing micro-organisms. With these, the causal agents of toxic as opposed to infectious types of food poisoning, the demonstration by serological or animal-challenge tests of toxins in suspect foods together with clinical symptoms of the patient (Table 9.2) provides sufficient evidence to confirm the cause of food poisoning. Indeed the organisms may well have died out in a suspect food and their toxins or some other metabolic product, for example the heat-stable DNAase of *Staphylococcus aureus*, may be the only evidence of a food having supported a large population of food-poisoning bacteria.

Table 9.3. Some properties of the toxins produced by food-poisoning bacteria.†

Bacteria	Toxin property
Staphylococcus aureus	Proteins, single-chain and globular; molecular weight, M_r 28 000–35 000. At least 7 (A, B, C, D, E & F) serologically distinct toxins recognized. Although termed enterotoxins, they are probably neurotoxins; they are absorbed from the gut, activate receptors on the abdominal viscera, the stimulus reaching the vomiting centre via the vagus nerve
Clostridium botulinum	Proteins, single polypeptide chains of *c.* M_r 150 000, which are 'nicked' by proteases when released from the bacterial cell; fragment A (e.g. M_r 50 000) is attached via a disulphide bond to fragment B (*c.* M_r 100 000). Six serologically distinct (A–F) toxins formed. The toxin probably associated with a haemagglutin (*c.* M_r 500 000) and is thereby protected from proteolysis and acid denaturation in the gut. Neurotoxin which is absorbed from the gut and binds, via gangliosides, to acetylcholine-containing nerve endings. Fragment B is probably the portion that binds to gangliosides and A is responsible for neurotoxic effects. These have not been fully defined; may block release of acetylcholine from nerve endings
Clostridium perfringens★	Protein (molecular weight, M_r 35 000) which is a structural component of the spore. Mode of action not known; the net flow of Na^+, Cl^- in the gut is reversed from absorption to secretion but not as a consequence of influencing intracellular levels of cyclic AMP (cf. Fig. 9.1), more probably due to the toxin acting on the membranes of cells of the gut epithelium
Bacillus cereus	Two or more toxins involved in food poisoning, one causing diarrhoea and an emetic (vomit-inducing) toxin suspected to be the cause of the two types of food poisoning which this organism produces (Table 9.2). The latter toxin is small ($< M_r$ 5000), heat-stable and probably spore-associated
Escherichia coli	Enteropathogenic strains produce two enterotoxins, one heat-labile (LT) the other heat-stable (ST). Both are coded on plasmid DNA. LT possess ADP-ribosylating activity and its mode of action mimics that of the cholera toxin (Fig. 9.2); antibodies formed against LT neutralize the cholera toxin, and vice versa. ST, a non-antigenic toxin of molecular weight M_r 4500–5000, which elevates the intracellular levels of cyclic GMP rather than cyclic AMP. Possible that ST induces diarrhoea by activating guanylate cyclase in the mucosa of the intestine

★ The account given in the text (p. 212) of 'pig bel', a major public health problem in Papua, New Guinea, identifies toxins in addition to that noted above.
† The relative toxicity of microbial toxins is shown in Fig. 9.1.

Although emphasis is given to the toxins (Table 9.3) produced by the causative agents of the toxic types of food poisoning, a satisfactory description of the symptoms caused by the infectious types also must rest ultimately on the identification of specific chemicals of microbial origin which cause demonstrable and interpretable perturbations in the biochemical activities of a target cell in or taken from a susceptible host. This situation obtains with food-poisoning strains of *Escherichia coli* in which plasmids code the synthesis of two toxins (Table 9.3), one heat-stable the other heat-labile. The latter shares certain properties, especially the stimulation of adenylate cyclase activity in the epithelial cells of the intestinal mucosa (Fig. 9.2), with *Vibrio cholerae*, the water-borne agent of cholera.

In practice, of course, the food microbiologist has to take a broad view of the microbiology of food poisoning and his major concern must be with practices that restrict contamination of products with food poisoning organisms and curtails growth of such organisms should contamination occur. Thus his/her role is that of a fire prevention officer rather than that of a fireman at the scene of a fire, the *modus operandi* of those faced with the investigation of outbreaks of food poisoning. In order to formulate sensible policies, attention has to be given to the following:

1   depots of infection;
2   routes of infection;
3   the resistance of food-poisoning organisms to adverse conditions;
4   the growth requirements of food-poisoning organisms;
5   a code of practice that minimizes contamination of foods and the growth of microorganisms;
6   refined and reliable methods for the quantitative recovery of food-poisoning organisms and their products;
7   a sampling plan commensurate with the degree of risk of a particular food.

(1) Depots of infection

Table 9.4 lists the principal depots of infection of the food-poisoning organisms. It will be appreciated that in any large food factory the range of products will most likely contain ingredients that link the factory to every depot listed in Table 9.4. Moreover, the factory's workforce must be regarded as the major depot of *Staphylococcus aureus* and, due to human carriers, an occasional depot of *Salmonella* spp. as well. Thus careful attention will have to be given to practices that are intended to screen manufactured products from contamination by the workforce. Some perspective can be given to what at first sight appears to be a complex situation by considering the range of foods that are commonly associated with outbreaks of particular types of food poisoning (Table 9.5) and the prevalence of different types of food poisoning in different parts of the world (Table 9.6). It must be appreciated that the information presented in Table 9.6 reflects features of a country other than the actual levels of particular food-poisoning organisms in the general, food factory or home

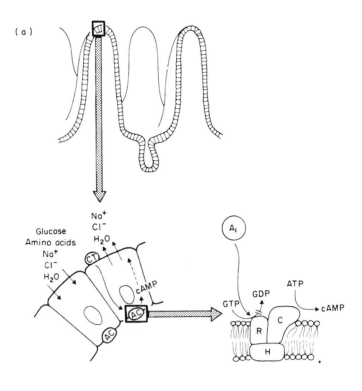

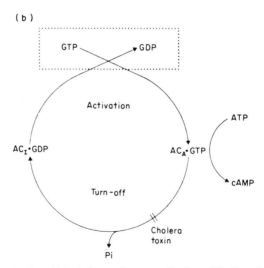

Fig. 9.2. (a) Mechanism by which cholera toxin causes diarrhoea. Binding of toxin to receptors on the lumen surface of ileal mucosal cells is followed by entry of fragment A_1, which interacts with the adenylate cyclase complex on the basal membrane, inhibiting the GTPase-mediated turn-off, the cyclase (probably by ADP-ribosylation of the GTP-dependent regulator protein). Increased intracellular cyclic AMP levels cause, by some as yet unknown mechanism, efflux of Na^+ and Cl^- ions, and hence also water. (b) Proposed mechanism for the regulation of adenylate cyclase activity and site of action of cholera toxin (II). AC_A: activated adenylate cyclase; AC_I: inactive adenylate cyclase.

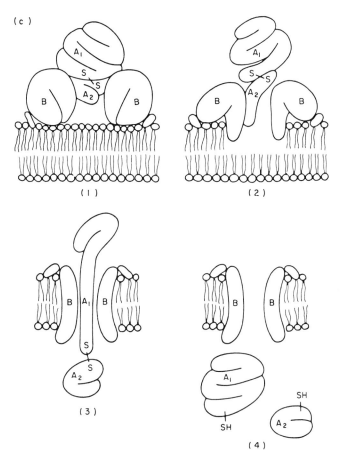

Fig. 9.2. (c) Proposed mechanism of entry of cholera toxin fragment A_I. Binding of B subunits to specific receptors (1) induces a conformational change in these subunits and their insertion into the membrane (2) to create a hydrophilic channel through which the A subunit can diffuse into the cell (3). Here, thiol agents reduce the disulphide bond, allowing fragment A_I to diffuse into the cytoplasm and activate adenylate cyclase. ((a and b) Redrawn from Stephen & Pietrowski (1981) *Bacterial Toxins.* © Van Nostrand Reinhold (UK) Co. Ltd.) ((c) Redrawn from Gill D.M. (1976) *Biochemistry* **15**, 1242–8. © American Chemical Society.)

environment. Thus the cuisine of a nation, or ethnic groups thereof, the extent of canteen or 'fast-food' catering and the medical resources devoted to public health will all influence annual statistics relating the incidence of food poisoning to particular organisms—in almost every country, the real incidence of most of these organisms in food poisoning can only be guessed at. The pre-eminence of *Vibrio parahaemolyticus* in Japan is correlated with the widespread use of raw fish and shellfish. The fish acquire the organisms from sea-water in the warm season (May–October); they can also transmit the organism to the home and cross-contaminate salted vegetables and cooked foods. Likewise a low but persistent incidence of botulism in the USA is commonly associated with the growth of *Clostridium botulinum* on soil-contaminated, low-acid vegetables (e.g. green beans) that have been canned

Table 9.4. The depots of bacteria that cause food poisoning.

Infectious types	Toxic types
Salmonella spp. Intensively managed farm animals, poultry and their environs. Humans, especially carriers	*Staphylococcus aureus* Humans, farm and domestic animals
Vibrio parahaemolyticus A marine organism of world-wide distribution, especially common in coastal and estuarine waters during summer months. Probably overwinters in sediments	*Clostridium botulinum* The majority of serotypes (Table 9.3) occur in soil; the psychrotrophic serotype E appears to be a common contaminant of water, sea sediments and mud
Escherichia coli Alimentary tract of man, farm animals and domestic pets	*Clostridium perfringens* The alimentary canal of man, farm animals and domestic pets; present also in soil, dust, etc.
Campylobacter jejuni Primary depot not identified	*Bacillus cereus* Common in soil and on vegetation
Yersinia enterocolitica Primary reservoir not identified	

in the home under conditions that cannot be relied upon to destroy the heat-resistant endospore. It is noteworthy, also, that the serotypes of *Clostridium botulinum* causing an outbreak almost always reflect the dominant serotype of the locality. Thus serotype B is common in the soils west of the Mississippi River and in foods that have become toxic whereas serotype A is most common in the east of America.

A list of foods (Table 9.7) that are rarely associated with food poisoning offers further help in identifying those that are especially prone to causing illness in susceptible persons. It has to be appreciated, however, that the good record of the former is normally dependent upon a process, like pasteurization, or composition, such as a high sugar content, that acts as a safeguard providing its effectiveness or presence is monitored routinely by the appropriate instrumental or chemical methods.

On some occasions, local or national emergencies may disrupt the processing of a food with the result that viable food-poisoning organisms are present in foods that have long been regarded as safe. An example of this situation is provided by an outbreak of *Campylobacter* infection in an area of Scotland. Severe winter storms caused a failure in the public electricity supply to a dairy. As a consequence, unpasteurized milk was distributed to towns and villages over an area of about 60 miles. Within 36 hours or so of this event, consumers, many of whom were from one to ten years of age, became ill—diarrhoea, abdominal pain, fever and headache being the principal symptoms. Over an eight-day period, about 148 persons became ill; the epidemic curve had a profile similar to that shown in Fig.

Table 9.5. Foods commonly associated with outbreaks of particular forms of bacterial food poisoning.

	Meat*	Poultry*	Fish*	Milk*	Eggs*	Sweets (puddings)	Vegetables*	Canned foods
Salmonella spp.†	+	+	±	+	+	+	−	±
Vibrio parahaemolyticus†	−	−	+	−	−	−	−	−
Escherichia coli	+	+	+	−	−	−	−	−
Yersinia enterocolitica	−	−	−	+	−	−	−	−
Staphylococcus aureus†	+	×	±	±	±	+	−	±
Clostridium botulinum	±	±	±	−	−	−	(+	+)
Bacillus cereus†	−	−	−	−	−	−	+	−
Clostridium perfringens	+	+	−	−	−	−	−	±

* and products thereof.
† tendency for increase in numbers of outbreaks during the warm season of the year.
+, common; ±, occasional; −, uncommon.

Table 9.6. Incidence of bacterial food poisoning in four countries (data, abstracted from Table VI in Epidemiology of food-borne diseases by F.L. Bryan (1979) in *Food-borne Infections and Intoxications* (Eds H. Riemann & F.L. Bryan) Academic Press, New York, refers to 1968–1972).

	England and Wales	USA	Japan
Salmonella spp.	[3149/11 041]†	185/10 329	414/—
Vibrio parahaemolyticus	1/12	9/1071	[1679/—]
Escherichia coli	—/—	6/1065	122/—
Yersinia enterocolitica	—/—	—/—	—/—
Staphylococcus aureus	96/1599	[208/12 562]	891/—
Clostridium botulinum	—/—	28/71	15/—
Bacillus cereus	6/31	6/126	—/—
Clostridium perfringens	187/6630	[102/25 356]	—/—

† Number of outbreaks/number of cases; square bracket indicates most prevalent causative organism in the particular countries.

Table 9.7. Processes or compositional factors that contribute to the public health safety of foods.

Process or compositional factor	Examples
Pasteurization, with protection from recontamination	Milk, liquid egg, ice cream
Appertization, with protection from recontamination	Canned foods
Fermentation, providing it is rapid	Sauerkraut
Low water content	Bread, biscuits and dried foods in general
High sugar content	Jam, fondant filling of sweets
High acid content	Pickles, non-alcoholic beverages

9.11. *Campylobacter jejuni* was isolated from the stools of 148 persons and 57 symtomless consumers of the unpasteurized milk. Attempts to identify the primary depot of infection were not successful and the investigators surmised that contamination of milk with cows' faeces may have been involved.

(2) Routes of infection

If a food-poisoning organism is a normal resident of the general environment in which a crop or an animal grows, then there is a high probability that harvested material or the animal will be contaminated by chance with low numbers of the organism. The actual levels of contamination may show seasonal variation, as with *Vibrio parahaemolyticus* on fish (see p. 179). This general situation would be expected to obtain with contamination of plant and animal material with *Bacillus cereus*, *Clostridium botulinum*, *Clostridium perfringens* and *V. parahaemolyticus*, all natural residents of soil or water. Likewise endospores of *Cl. botulinum* are occasionally found in pork. With these examples, a direct route of infection would appear to operate. Any additional contamination would be the consequence of a product's exposure to dirt or dust during storage. The actual level of the initial contamination would be accentuated if conditions conducive to the growth of food-poisoning organisms obtained during the storage and processing of a commodity.

Even when a food-poisoning organism is not a common resident of arable or grazing land, contamination of a crop or animals with it could result from farm husbandry systems. Thus the use of faecally contaminated water to irrigate salad crops might be expected to contaminate plants with organisms, including *Salmonella* spp., of gut origin. Indeed it was noted in the discussion of water microbiology (see p. 142) that the Bournemouth/Poole outbreak of typhoid fever was attributed to the contamination of cows' udders with water polluted by a human carrier. Moreover, there is evidence that unchlorinated effluent water from sewage plants using either biological filters or the activated sludge system can contain *Salmonella* spp. Thus the use of such water for irrigation purposes would have an element of risk attached. The disposal of sludge or, more especially, the wastes from intensively housed animals, pigs or poultry on to farmland must also be considered as a means whereby crops or grazing animals could be contaminated with *Salmonella* spp. Birds can be involved in the dissemination of salmonellae; for example those that spend a part of the day on a municipal dump, a sewage plant, or an effluent-pipe from it, and the other part on farmlands have a life style that might be expected to transmit pathogenic organisms from one site to another. Indeed circumstantial evidence led recent investigators to this conclusion when they isolated an unusual serotype, *Salmonella zanzibar*, from the bulk milk tank on a Scottish farm (Fig. 9.3). This episode was interesting also in that it is probable that the organism was brought to Scotland from Malaysia by a holiday-maker.

In practice such transmission routes are probably of minor concern only when compared with those that can occur in agricultural systems based on intensive husbandry of calves, pigs or poultry. The potential depots of infection and transmission routes for salmonellae within the poultry industry are shown in Fig. 9.4. This Figure indicates that, at a conservative estimate, a food factory using shell eggs and egg and poultry products would be at the end of a chain having six potential depots of infection (Depots 1, 2, 3, 4A, B and C of Fig. 9.4). If the factory operated a policy of taking back products from retailers or the housewife (see p. 65), then it would be linked with a further three depots (6A, B and 7). In addition the factory could be linked, albeit indirectly, with Depot 8 if, as would be probable, the factory had waste materials collected by firms that process swill for feeding to pigs or that recover protein for animal feeds. The last-mentioned operation

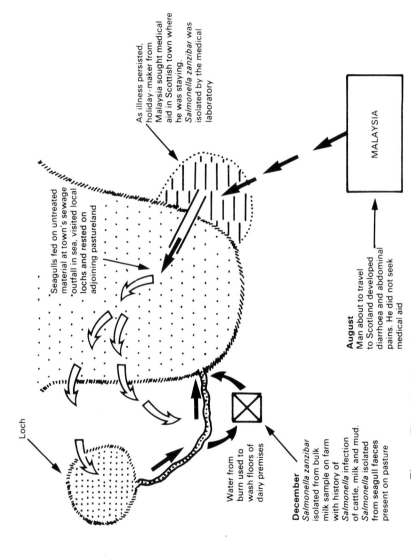

As illness persisted, holiday-maker from Malaysia sought medical aid in Scottish town where he was staying. *Salmonella zanzibar* was isolated by the medical laboratory

MALAYSIA

August
Man about to travel to Scotland developed diarrhoea and abdominal pains. He did not seek medical aid

Seagulls fed on untreated material at town's sewage outfall in sea, visited local lochs and rested on adjoining pastureland

Loch

Water from burn used to wash floors of dairy premises

December
Salmonella zanzibar isolated from bulk milk sample on farm with history of *Salmonella* infection of cattle, milk and mud. *Salmonella* isolated from seagull faeces present on pasture

Fig. 9.3. The chain of events that was responsible for transferring *Salmonella zanzibar* from Malaysia to the bulk milk tank on a farm in the north of Scotland.

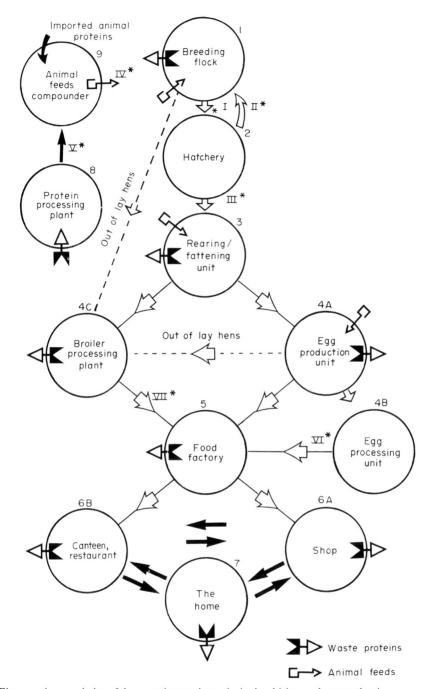

Fig. 9.4. A general plan of the many interacting units in the chicken and egg production industry.* Control points: I, fumigation with formaldehyde or disinfection of the shells of hatching eggs with gluteraldehyde; II and III colonization of the caeca of newly hatched chicks with anaerobic bacteria (the Nurmi process), IV disinfection of compounded feeds or feed additives, V disinfection of protein wastes, and VI pasteurization of egg products.

would have the potential to introduce salmonellae present on scraps from a food factory back into Depots 1, 3 and 4A of Fig. 9.4, either directly or indirectly through Depot 9. Moreover, the possibility exists on a farm for cross-contamination between pigs and poultry, either through contamination of the environment or the use of common equipment for compounding animal rations. Indeed the latter operation could link a farm to depots of salmonellae contamination other than 8 and 9 of Fig. 9.4. Thus the importation of protein supplements from other countries can link the farming operations of one country with those of another or a particular industry of the exporting country, for example the manufacturer of fish meal for use in animal feeds. In the flow diagram in Fig. 9.4, members of the workforce of any of the enterprises could be carriers, either chronic or temporal, and thereby introduce salmonellae. Thus a diagram such as this can be regarded as an oversimplified view of the transmission routes and perhaps it would be best to consider the transmission of salmonellae within and between facets of agriculture and food processing as a web rather than simple steps from one depot to another.

It is probably unrealistic to contemplate the eradication of salmonellae from intensively managed farm animals. In practice one has urbanized man linked directly to 'urbanized' animals; the former are serviced with sophisticated systems of medicine, refuse and sewage disposal, whereas the latter lacks such support other than during times of emergency. Although eradication may be impracticable, controls designed to minimize cross-contamination and to limit the extent of contamination must be considered. The following discussion will be concerned with control systems that may be applied to a simplified overview of salmonella transmission within the poultry industry (Fig. 9.5).

(a) POULTRY FEEDS

Evidence from epidemiological studies in the United Kingdom has shown that *Salmonella* spp. imported in fish meal or animal protein meal have been introduced via animal feeds to breeding units of poultry and turkey. Within a short time of introduction, human disease has been caused by the imported serotype. Over a longer time scale, scattered human outbreaks have been linked back via distribution/retail networks to poultry processing plants that receive broilers or turkeys from infected flocks. (An example of the widespread dissemination of a food-poisoning organism, *Escherichia coli*, by a food commodity is discussed on p. 192.) This situation highlights two important features of food poisoning due to salmonellae. In the United Kingdom, for instance, *Salmonella typhimurium* was, until recently, the major cause of food poisoning. Indeed the sum of the outbreaks it caused exceeded the sum of outbreaks due to all other serotypes. Within the past couple of years the position has been reversed. When serotypes other than *typhimurium* are considered, the records show that within a country there are 10–20 predominant ones. Moreover, records relating to several years show

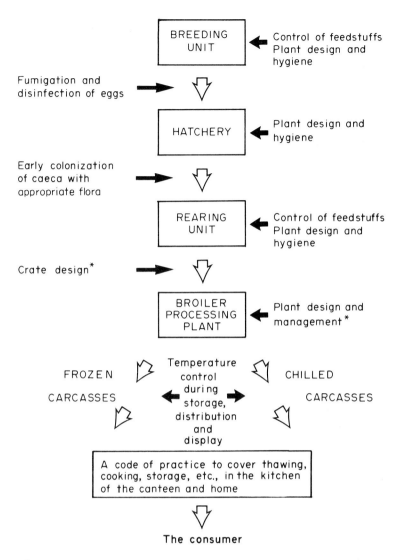

that different serotypes contribute to the 'top ten' at different times. In the UK, some of these organisms are known to have been introduced in imported proteins used for animal feed. They have not achieved, however, a predominant position as a consequence of repeated introductions into the country but through becoming established in poultry flocks or their environs. Thus each depot identified in Fig. 9.4 has the potential to perpetuate a clone of a particular serotype of *Salmonella*. There are several practical approaches that may be considered to prevent introduction (applied at asterisks III, IV and V, Fig. 9.4):

1   treatment of compounded feeds;

2 treatment of proteins and other wastes recovered from food factories, poultry processing plants and abattoirs for conversion into bone meal, meat meal, feather meal, dried blood, mixtures thereof or swill.

Before discussing various treatment methods, the scale of the problem facing the poultry industry must be considered. As noted above, epidemiological evidence has linked outbreaks of food poisoning in humans with *Salmonella* serotypes introduced by protein supplements in feeds to turkey and poultry flocks. Does an outbreak follow every introduction? The results of two major surveys suggest that this is not the case. Thus a British survey of 4140 samples found salmonellae in 9% of raw ingredients, 2.8% of finished meals and 0.3% of pelleted feeds; 44 *Salmonella* serotypes were isolated. An American survey of 5712 samples of animal feed revealed that 12.6% contained salmonellae and 59 serotypes were identified. Such evidence of a low but by no means negligible level of contamination suggests that perhaps only occasionally does a serotype become lodged in a poultry unit. Moreover, the number of organisms per unit weight of feed probably determines to some extent whether or not salmonellae become established in a flock or its environs. As only 1000 *Salmonella montevideo* per gram of food can cause infection of turkeys, an infectious dose much smaller than that required to infect humans (cf. Table 9.8), the extent as well as the incidence of salmonellae contamination of feedstuffs is important also but there appears to be relatively little information on the former. The evidence on the incidence of contamination leaves little doubt, however, about the need to consider methods of treating protein supplements and compounded feeds for poultry.

Table 9.8. Infectious doses of food-poisoning bacteria.

Bacteria	Dose
Salmonella spp.	$> 10^5$ organisms in food ingested by susceptible host*
Clostridium perfringens	10^8–10^9 organisms in food ingested
Staphylococcus aureus	1 μm toxin usually $> 5 \times 10^6$ organisms present in food ingested by susceptible host
Bacillus cereus	$> 10^5$ organisms in food ingested

* Many fewer *Salmonella* appeared to cause illness in a recent outbreak, see Gill *et al.* (1983) The Lancet, i, 575–577.

Various approaches have been taken in attempts to rid ingredients or poultry feeds of *Salmonella*: irradiation, addition of antimicrobial agents, heat treatments or combinations of these. With irradiation, doses of 0.5–1.0 Mrad (see p. 81) have been shown to produce five or more decimal reductions in the number of viable salmonellae in experimentally contaminated feed without demonstrable damage to lysine, an essential amino acid in poultry nutrition. Such treatments do, however, have the potential to accentuate oxidative rancidity of feeds. Formaldehyde can decontaminate animal feeds but its commercial application could be limited by technical problems associated with diffusion of gas through a finely powdered meal. A similar problem would appear to obtain with ethylene oxide, a bactericidal agent that has been assessed for its potential to reduce salmonella contamination of meat and bone meal. Methyl bromide, propionic acid and fatty acids in general have been assayed also for use as sterilizers of poultry feed. The overall experience of persons who have attempted to diminish the level of contamination of feeds with salmonellae suggests that pasteurization would probably be the easiest system to apply. A great range of temperatures and times of application are given in the literature. This situation is due in part to the methods used to apply the heat (either during the production of pellets or during expansion and extrusion of feed) and in part to the moisture content of the feed. Indeed it would seem that an acceptable method will not be available until a thorough systematic study establishes time–temperature relations that need to be applied to feeds of different moisture contents. With swill, boiling or retorting could be expected to eliminate salmonellae. It must be appreciated, however, that whatever methods are adopted they will have to be applied together with systems that preclude recontamination of treated materials with untreated ingredients, a major problem in an industry where recontamination by dust is likely.

(b) TRANSMISSION OF SALMONELLAE

The above discussion dealt with possible means of controlling the introduction of salmonellae to poultry rearing or breeding units by one vector, animal feeds. If control were to be achieved, little of value would result if attempts were not made also to prevent, or at best impede, transmission of salmonellae from one stage to the next in the chain starting with the fertile egg and terminating with the broiler chicken ready for use in the home or canteen. Possible checks to transmission (Fig. 9.5) are:

1 freeing the hatching egg of salmonellae;
2 attempting to impede colonization of the chick with salmonellae;
3 adopting production systems that prevent or at best minimize the transfer of salmonellae from the environment to the bird, from bird to bird, carcass to carcass, or poultry product to raw materials or other prepared foods, especially cooked ones, in the kitchen of the home or canteen.

It will be appreciated that there are problems, both statistical (how would you obtain a random sample of eggs from a breeding unit producing 1000 eggs per day?) and sampling (having randomly selected your sample, how would you maintain asepsis when analysing the viscous albumen?), which have yet to be overcome before a reliable figure on the incidence of salmonellae infection of eggs can be obtained. The best estimate to date is that 1 in 1–2000 turkey eggs may harbour salmonellae and that very many fewer eggs contain salmonellae in the yolk or white at the time of oviposition. Thus attention is directed at contamination of the shell and the means whereby this can be eliminated. As the nest boxes are the initial and probably the most important source of infection because salmonellae-infected faecal material may be introduced on the feet and plumage of hens, current research seeks to devise systems that protect the egg from contamination. Some success has been achieved by adding paraformaldehyde flakes to traditional nest litter (sawdust or chopped straw) or by replacing such litter with man-made materials such as Astroturf.

Attempts are being made also to free hatching eggs of salmonellae acquired through contact with contaminated materials. As the traditional method of egg fumigation with formaldehyde in the gaseous form has not been successful in preventing the dissemination of food-poisoning salmonellae, current investigations are concerned with procedures that could be used to deposit antimicrobial agents in the cuticle, on the wall of the pore canal and in the shell membranes (Fig. 4.7) without adversely affecting hatchability. To ensure that antimicrobial agents reach the sites noted above, techniques diametrically opposed to those used to wash table eggs (see p. 82) are being examined—warm eggs are allowed to contract in cold solutions of disinfectants. As yet, however, there are no results from field trials with which to assess the efficacy of this approach to egg disinfection.

The caecal sacs of young chicks are rapidly colonized with salmonellae when these organisms are present in the environment in which the chicks are held for a day or so after hatching. As the caeca of mature fowls contain a resident flora of facultative, obligate and strict anaerobes, for example *Bacterioides* spp., much current research effort is being devoted to systems that ensure rapid colonization of the chick's caeca with a flora that excludes salmonellae. Promising results were obtained by the pioneer of this system, Professor Nurmi who used caecal contents of hens to infect newly hatched chicks. Results justifying further surveys have been obtained from experiments with a 'cocktail' of organisms isolated from the caeca, grown in pure culture and mixed immediately before administration. It will be appreciated that the time of administering the 'cocktail' is important— salmonellae-infected caeca are not rapidly recolonized by members of the cocktail—and that routine dosing of hundreds, if not thousands, of newly hatched chicks with fastidious organisms under farm conditions could well pose technical problems.

(c) TRANSFER OF SALMONELLAE

If either of the methods discussed above proves effective, then it is imperative that the buildings used to house salmonellae-free chicks do not cause recontamination. As low-cost materials are commonly used to construct houses for intensive animal husbandry, effective cleaning and disinfection may pose problems simply because penetration of steam or disinfectants into nooks and crannies cannot be assured. Moreover, a rigorous programme of disinfestation needs to be linked to cleaning procedures so that salmonellae harboured by the larval stage of insects, such as flour moths, are not reintroduced into the building.

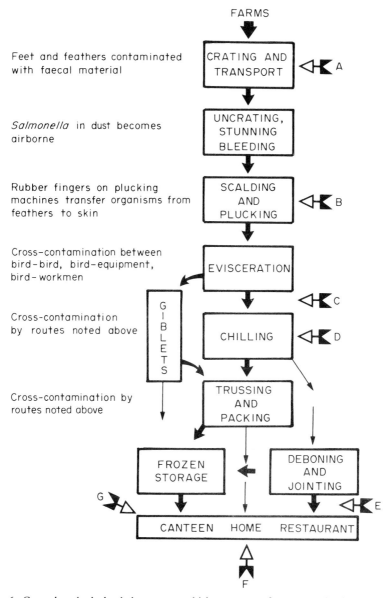

Fig. 9.6. Control methods that help to protect chicken carcasses from contamination with *Salmonella*. The letters A–G refer to Table 9.9.

Figure 9.6 directs attention at potential causes of cross-contamination in a broiler processing plant and possible remedial measures are listed in Table 9.9. It will have become apparent from the discussion in this section that no single method of control can be relied upon to impede the transmission of salmonellae through an intensive system of animal production and large-scale processing of the final product. Instead, several methods of control have to be considered for each major link in the production chain,

Table 9.9. Some possible approaches to the control of salmonellae contamination of poultry carcasses.

Process*	Means of controlling contamination
A Crate design	A solid floor prevents the faeces from one layer of birds dropping on those in a lower layer when crates are stacked on a lorry
B Scalding	An alkaline reaction (pH 9.0 with sodium carbonate) and heating at 58 °C helps maintain low levels of microorganisms in the scald tank
C Spraying	Washing carcasses with a spray of water helps reduce the level of contamination
D Chlorination and water flow	An adequate flow rate of chlorinated water contributes to low levels of contamination of carcases (Fig. 5.3)
E Efficient refrigeration	Storage at 0–4 °C prevents growth of salmonellae
F Effective thawing	Packaging materials ought to provide clear instructions on the preferred means of thawing deep frozen poultry under conditions that will not encourage growth of salmonellae
G Education	The training of catering workers ought to give emphasis to (i) the need to keep raw and cooked foods apart; and (ii) the need to prevent contamination of cooked foods with equipment used in the preparation of uncooked material

* See Fig. 9.6.

as indicated in Fig. 9.5, and the relative effectiveness of each considered so that optimal value is obtained from the limited resources of manpower and equipment available for routine monitoring. Indeed there is probably a need to contemplate mathematical modelling of an integrated system such as that depicted in Fig. 9.5, so that the important vehicles of introduction of a food-poisoning organism as well as the major causes of transmission can be identified by risk factor analysis.

(d) AN EPIDEMIC DUE TO *ESCHERICHIA COLI* 0124

It was noted on p. 186 that sporadic outbreaks of food poisoning due to particular serotypes of *Salmonella* have been associated with the establishment of the organisms in poultry breeding or rearing units. The sporadic nature and widespread distribution can be attributed to two factors; the relatively low levels of contamination of chicken or turkey carcasses and the common use of deep-freezing for storage and distribution. The development of populations of sufficient size to cause illness in a person (Table 9.8) is generally associated with unsatisfactory conditions for thawing or cooking this commodity. The following account of an epidemic of infectious-type food poisoning due to *Escherichia coli* 0124 directs attention at problems that can arise from process-contamination of a product that is not deep-frozen during storage and distribution or cooked before use.

Between 29 September and 14 November 1971, 2400 8 oz packs of soft, ripened cheese were imported into the USA. In the period 30 October–10 December 1971, 107 episodes—of which 96 occurred in the last 2 weeks of November—of gastroenteritis in at least 12 states were associated with this product. In 105 of these episodes, 387 persons (attack rate, 94%) became ill. It was noted that all the cheese associated with the confirmed cases of gastroenteritis was manufactured in a 2-day period at one dairy in the exporting country and it was shown that samples of the cheese produced at that time contained $c.$ 10^5–10^7 $Escherichia$ $coli$ 0124 g^{-1}. The source of contamination of the cheese was not identified. This epidemic highlights the high incidence and widespread distribution of illness that can be associated with in-plant contamination of a product with a causative agent of the infectious type of food-poisoning microorganism.

Botulism

Attention was given in the above discussions to vectors that have the potential to transmit a food-poisoning organism along a food chain to a product that even if cooked sufficiently to kill salmonellae may still be recontaminated by carriers or equipment. If an adequate time interval and storage temperature (Table 9.10) obtain post recontamination, then a population of salmonellae sufficient in numbers (Table 9.8) to cause food poisoning could be consumed by a susceptible person. With canned foods, the emphasis given to retorting (see p. 76) and asepsis (see p. 84) ought to, in theory, ensure that any probability of toxin production by $Clostridium$ $botulinum$, for example, is extremely remote. This situation obtains in general but the occasional occurrence of botulism following the consumption of a canned food warns against complacency and the dangers of 'tunnel vision' directing attention exclusively at asepsis as influenced by can construction (Fig. 4.7), the retorting operation and post-process handling. An outbreak in Birmingham, UK, in 1978 drew attention to other factors.

Four elderly persons, two male and two female, took afternoon tea and within 14 hours they were admitted to hospital with symptoms of botulism. Intraperitoneal injection of mice with 0.5 ml of a patient's serum led to the death of those that had not been protected by antiserum against $Clostridium$ $botulinum$ type E. The same results were obtained with mice injected with portions of 20 ml of 0.85% saline which had been used to swill out the can that had contained salmon consumed at the tea party. $Cl.$ $botulinum$ type E was isolated in pure culture from salmon remnants in the can. The can, which had been filled and retorted one year before in another continent, had a small area of damage of unknown cause. Circumstantial evidence led to the conclusion that spores of $Cl.$ $botulinum$ type E had penetrated through the damaged area post retorting; they germinated and the vegetative cells formed toxin in the year during which the can was stored. Moreover, as the production line at the factory was effectively U-shaped,

Table 9.10. Growth-limiting factors for some food-poisoning bacteria.

Bacteria	Growth-limiting factors
Salmonella spp.	Temperature range, 5.7–45.6 °C pH limits, 4.5 and <9.0 a_w limits, 0.93 Killed, by mild heat, e.g. 60 °C slowly
Clostridium botulinum (other than the psychrotrophic type E and non-proteolytic type B)	Temperature range, 10–50 °C Oxygen is toxic pH limits, 4.5 and 9.0 a_w limits, 0.95 Spores killed slowly at 100 °C
Clostridium perfringens	Temperature range, 6.5–53 °C Oxygen can be toxic pH limits, 5.0–9.0 a_w limits, 0.95 Spores killed by temperatures of 100 °C+ but marked variation between strains
Staphylococcus aureus	Temperature range, 7.8–45.6 °C pH limits, 4.5 and <9.0 a_w limit, 0.86 (aerobic), 0.90 (anaerobic) Killed slowly at 60 °C
Bacillus cereus	Temperature range, 7–49 °C pH limits, 4.9, (?) a_w limit, 0.96 Killed slowly at 100 °C

there would appear to have been an opportunity of processed cans coming into close proximity with raw materials, contaminated equipment and the workmen who gutted salmon. Moreover, as intoxication of the can of salmon resulted from the growth of a saccharolytic, non proteolytic *Cl. botulinum* type E, there were no overt signs of spoilage to alert the person who opened it. A proteolytic strain would have made the fish unpalatable and gas produced from the breakdown of carbohydrates had escaped via the damaged area thereby preventing a build up in pressure and the consequent swelling of the can. Several features of this outbreak are worthy of note: the thousands of miles that separated the site of infection of the can and the location of the outbreak; a defect that would go undetected if attention were paid only to effective working of the can 'sealer'; poor design in the layout of a processing line, and toxicity of the salmon without overt signs of spoilage.

(3) The resistance of food-poisoning bacteria to adverse conditions

Although there have been many studies of the resistance of food-poisoning bacteria to heat, a_w and pH, a summary (Table 9.10) of the major observations has to be used judiciously. There is a latent danger in taking a statistic from such a summary and applying it generally. Thus even with a cardinal point on an environmental gradient (see p. 3), one must be alert to the possibility that a change in formulation or the method of producing a food might conceivably override the supposed safeguard before laboratory studies had warned of the probability of such an event occurring. Thus with pH, a value of 4.6 has long been held to be that at which *Clostridium botulinum* will not grow. It has been reported recently, however, that it will grow under slightly more acid conditions when certain acids (i.e. citric or hydrochloric) are used to poise the pH of a culture medium. Moreover, problems could arise if emphasis was given to a cardinal point without considering also other features of a food or the possible abuses to which it might be exposed during storage and distribution. Thus if metabolizable organic acids contributed significantly to pH 4.6 in a food containing viable organisms, the latter might grow at the expense of these acids during periods of temperature abuse thus causing an alkaline drift and paving the way for the eventual outgrowth of vegetative cells from the endospores of *Clostridium botulinum*.

It is evident from the above that any considerations of the possible role of the inimical factors, (a) high processing temperatures, (b) low a_w, (c) low pH and (d) low or high storage temperatures, must be set against the public health record of the particular food commodity in question. It was noted previously (see p. 72) that the 12-D treatment of low-acid foods produces commodities with exemplary public health records. Indeed when problems occur, they are often traced back to post-process contamination (e.g. the Aberdeen outbreak of typhoid fever, see p. 142) or can damage (e.g. the Birmingham outbreak of botulism, see p. 193). There are certain low-acid foods that receive less than a 12-D treatment but which in general have a good public health record. Canned sardines are one example— organisms surviving heat treatment are probably inhibited by a component of the olive oil. Canned cured-meats are another; because the safety of the latter is dependent upon the interplay of many factors, it is discussed in detail in a following section (*Clostridium botulinum*). That discussion may be considered also to foreshadow future laboratory studies and practical applications in which attempts are made to ensure the safety of a food by causing several inimical agents to interact rather than relying upon one agent alone (cf. Table 9.10). With high temperatures, for example, it is easy for the management of a food factory or canteen to assume that the safety of a product will be assured even though rigorous protection against re-contamination or control of the cooling phase is not exercised. The potential microbiological problems of such a situation are discussed on p. 198. Similarly with low-temperature (chill) storage, reliance upon unmonitored mechanical refrigeration is not advisable. Thus, for example, critical temperature control of a refrigerator would need to be assured before contemplating the chill storage of fish or fish products that may be contaminated with *Clostridium botulinum* type E, an organism that can grow at 3 °C.

(4) Growth requirements

Physical and chemical factors have to be considered in a discussion of the growth requirements of food-poisoning bacteria. It is easy to list factors (Table 9.10) that have been shown to influence the rate and extent of the growth of an organism and toxin production in the laboratory. In practice there are advantages to be gained on occasions by piecing together field observations of outbreaks of food poisoning and setting the evidence against information obtained from laboratory studies. Such an approach not only provides a useful perspective of information obtained with pure cultures but, through identifying commercial or catering practices that favour the growth of a particular organism, it ought to lead to codes of practice whereby effective control of a particular organism in a commodity or general class of commodities is achieved. To illustrate this approach, factors contributing to outbreaks of food poisoning due to *Clostridium perfringens* will be discussed.

CLOSTRIDIUM PERFRINGENS

A toxin formed (Table 9.3) during endospore formation by vegetative cells in the alimentary canal is the primary cause of the clinical symptoms (Table 9.2) of food poisoning due to *Clostridium perfringens* type A, an alpha-toxin (lecithinase)-producing organism of ubiquitous distribution in nature and the alimentary tract of man and animals. In order for a toxic dose to be produced, a person has to consume a relatively large number of vegetative cells—about 5.0×10^8 vegetative cells are considered to be a critical dose. Table 9.5 indicates that meat and poultry, especially those cooked for bulk catering, are the foods most commonly associated with this type of food poisoning. As Robertson's cooked meat medium is commonly used to grow clostridia in the laboratory without recourse to anaerobic jars or cabinets, it can be accepted that cooked meat and poultry will provide the essential nutrients for growth of *Cl. perfringens* and it can be assumed also that a satisfactory redox potential will obtain. Thus the emphasis can be placed on other environmental factors. An American survey of 39 outbreaks identified the following (listed in descending order of importance): improper holding temperatures, inadequate cooking, contaminated equipment and poor hygiene. With inadequate cooking, vegetative cells will grow during the heating phase, remain quiescent at the surface of the meat if and when the temperature goes above 60 °C and then resume growth during a period of uncontrolled cooling. Even with adequate cooking and a rapid rise in temperature during the heating phase, endospores may not be inactivated. Indeed the temperatures may be such that they are heat-shocked and thereby modified so that a rapid and high incidence of germination occurs once the temperature of the meat again enters the zone in which *Cl. perfringens* grow. In addition heating will also reduce the redox of the meat, improve its nutritional status through denaturing proteins and, through destroying the vegetative cells present on the raw meat, remove would-be competitors to the nascent vegetative cells of *Cl. perfringens*. Even with the bacterial endospores on the meat, heat could be expected to have a selective action—the most heat-resistant ones surviving for the longest time. Indeed the earliest studies of outbreaks of food poisoning due to *Cl. perfringens* drew attention to the exceptional heat resistance of the endospores—they survived 100 °C for 1 h whereas those of the 'classical' strains of this species were killed within 10 min at this temperature. Subsequent investigations have shown, however, that exceptional heat-resistance is not a unique feature of all the strains of *Cl. perfringens* isolated from foods that were known to be the cause of food poisoning.

As the above discussion has identified temperature as the important environmental factor in the events leading to outbreaks of food poisoning, it is pertinent to consider the growth rate of *Clostridium perfringens* at various temperatures. Extensive surveys of the growth of a variety of strains of this organism in a range of meat products have shown that the generation times (in minutes) can be as short as 8.5 in raw ground beef at 45 °C or as long as 39.4 in cooked ground beef at 26 °C or 30.8 in cooked ground beef at 51 °C. Growth has not been detected at temperatures above 53 °C and the available evidence suggests that, even with prolonged storage, growth does not occur at 15 °C or less. It is notable also that a lag phase of growth has not been noted with strains of *Cl. perfringens* in meat held at 46 °C whereas a 2–4 hour lag has been noted at 35 °C. Only an elementary knowledge of mathematics is needed to appreciate that very large populations of *Cl. perfringens* will develop in 2–4 hours if cooked meat contaminated with 1000 organisms having a generation time of 8.5 min at 45 °C is stored at this temperature. If the evidence relating generation times and temperature is considered in terms of the hypothetical cooling curve of cooked meat (Fig. 9.7), then it is obvious that the period (x–y) during which the meat

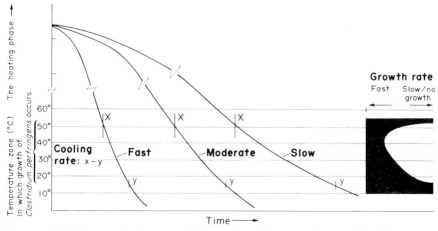

Fig. 9.7. The influence of cooling rate of a food on the relative growth rates of *Clostridium perfringens*.

is in the physiological temperature zone of *Cl. perfringens* will be short with a fast but long with a slow cooling rate. The implication of these hypothetical cooling rates must be considered in all operations in which the cooling of meat under conditions that do not ensure asepsis or additional heat treatment before consumption are contemplated. Thus temperatures above 60 °C ought to be selected if sliced cooked meat is likely to be stored for some appreciable time before serving. Moreover, the management of cooked meat must be based on the temperatures obtaining in the meat rather than the assumed temperature of the store. Thus it has been demonstrated that under exceptional circumstances the growth of *Cl. perfringens* can continue for up to 10 h in cooked meat or meat products in a refrigerator. Moreover, it must be recognized that fast cooling and effective refrigeration alone cannot be relied upon for the control of *Cl. perfringens* in a cooked product subjected to temperature abuse following refrigerated storage. Although refrigeration will arrest the growth of the organism and may even cause a slow decrease in the size of its populations, viable cells will resume growth if the meat is subsequently held in the range 15–53 °C (Fig. 9.7).

Although *Clostridium perfringens* has been singled out for discussion in this section, the principles which have been considered are relevant also in considerations of factors leading to food poisoning due to the other endospore-forming organisms such as *Bacillus cereus*.

BACILLUS CEREUS

This organism, which is ubiquitously distributed in nature, will be present occasionally in a range of foods but in such low numbers that there is little if any risk of illness to a consumer. If, however, the methods of preparing certain types of food (commonly those rich in carbohydrates) eliminate would-be competitors and provide both a suitable temperature and adequate time for microbial growth, then *Bacillus cereus* can form large populations (upwards of 10^8 organisms g^{-1} or ml^{-1}) and cause either the diarrhoeal or vomiting type of food poisoning (Table 9.2). An investigation of 110 incidents of the latter in the UK revealed that 108 were associated with cooked rice, one with pasteurized cream and the other with vanilla slices. Detailed surveys of the rice-linked incidents identified several practices that could be expected to (a) kill organisms other than spore formers, and (b) provide opportunities for the out-growth of endospores. Thus large amounts of rice may be cooked by boiling, allowed to 'dry off' at kitchen temperatures and be stored for a few hours or even days during which time portions are taken and 'flash fried' for customers' needs. As a result of these surveys the following code of practice was suggested.

(i) Rice should be boiled in small quantities—this reduces storage time before frying.

(ii) Boiled rice should either be kept hot or cooled quickly and transferred to a refrigerator within 2 h.

(iii) Boiled or fried rice must not be stored at temperatures between 15–50 °C.

It will be appreciated that this code is based on two premises: (1) endospores of *Bacillus cereus* in raw materials are unlikely to be eliminated by boiling, and (2) the methods of storage must not cause rice to be held for long periods at the growth temperature (7–49 °C) of the organism (Table 9.10).

YERSINIA ENTEROCOLITICA

Although a member of the family Enterobacteriaceae, commonly regarded as mesophilic bacteria, this species is capable of growing and forming large populations at 4 °C, a temperature that, with the exception of the psychrotrophic serotype of *Clostridium botulinum*, is often regarded to be one that will not support growth of food-poisoning bacteria.

This organism's role as an agent of food poisoning was established in 1976 following a thorough investigation of an outbreak of abdominal illness characterized by pain, diarrhoea, fever and, in some cases, rash. The symptoms were so acute in 36 children that they were admitted to hospital and 16 of them had appendectomies. Epidemiological studies implicated chocolate milk, which was sold in the cafeterias of the schools attended by the children, as the mediating agent and led the investigators to the dairy that produced the drink. In the dairy, chocolate syrup was added to milk which had been pasteurized, the actual mixing being done 'by hand with a perforated metal stirring rod' in an open vat. *Yersinia enterocolitica* was not isolated from the pasteurized milk but it was present in one carton of chocolate milk. This evidence led to the suggestion that the pathogen may well have been introduced into the milk at the mixing stage in the dairy.

CLOSTRIDIUM BOTULINUM

It was noted at the beginning of section (4) that there were advantages to be gained by considering field observations alongside evidence from laboratory study of a causal organism of food poisoning. This situation obtains only if the practical importance of a nutritional or environmental factor can be identified with certainty, as is the case generally with temperature and food poisoning due to *Clostridium perfringens*, especially as the consumption of a large number of vegetative cells is a prerequisite of an outbreak. When technological modifications of the processes used to preserve a food with a good public health record need to be considered because of changes in the demands of the public or clinical evidence that a traditional food may pose a threat other than a microbiological one to the health of a consumer, then field observations are likely to be of limited value especially if the public health record of a product reflects the interplay of several factors. Moreover, if toxin production in a food is a prerequisite of food poisoning, monitoring of vegetative growth alone may provide inadequate information. Recent studies of *Cl. botulinum* illustrate the various issues raised above.

Public health records indicate that meats 'cured' commercially with NO_3^-, NO_2^- and salt, with or without smoking, are rarely implicated in botulism. Moreover, cured meats that have been canned and given little other than a pasteurizing treatment also have a good public health record. Evidence obtained in the last 20 years has shown that cured meats may contain nitrosamines, especially if subjected to high temperatures when cooked (see p. 40).

A summary (Fig. 9.8) of the technological changes in the American meat industry in the past 80 years draws attention to two innovations (the adoption of O_2-impermeable wrapping materials, and the use of mild heat treatment for canning hams and other cured meat products) and three major trends:

(a) a progressive increase in the use of mechanical refrigeration and the establishment of an effective 'cold chain' between the meat packing plant and the home or canteen;

(b) a gradual elucidation of the roles of the various agents used in meat curing;

(c) the realization that nitrite might at times react with meat components to produce nitrosamines.

Although the available literature favours a discussion of the American experiences, these are not unique to that country. Moreover, it will be evident in the following discussion that the convenience of identifying (a), (b), (c) separately must not be allowed to obscure the interactions between them.

(a) *Wrapping materials*

The adoption of O_2-impermeable films for packing hams and bacons caused some concern because of fears that their beneficial effects (hindering the rate of fading of the cured meat colour and growth of aerobic spoilage organisms) might be offset by an increased risk of botulism, the growth of *Clostridium botulinum* being favoured by anoxia. Although the public health records for the past 25 years reveal that these fears were unfounded, experiences in the fish industry showed that they were warranted at the time. Thus within 3 years of the introduction (in 1960) of vacuum-packed smoked fish, there were 21 cases and 9 deaths caused by the consumption of fish containing toxin produced by the psychrotrophic (grows at $3\,°C$) *Cl. botulinum*. This position has been attributed to the failure of the industry to inform traders and consumers of the potential dangers of a novel product or to offer advice on storage conditions.

The use of impermeable films for wrapping cured meats might be expected also to favour the growth of *Staphylococcus aureus*, an organism that grows well in the absence of oxygen and in the presence of salt. Its failure to form populations of appreciable size is probably due to its inability to compete with the resident flora especially with chill storage.

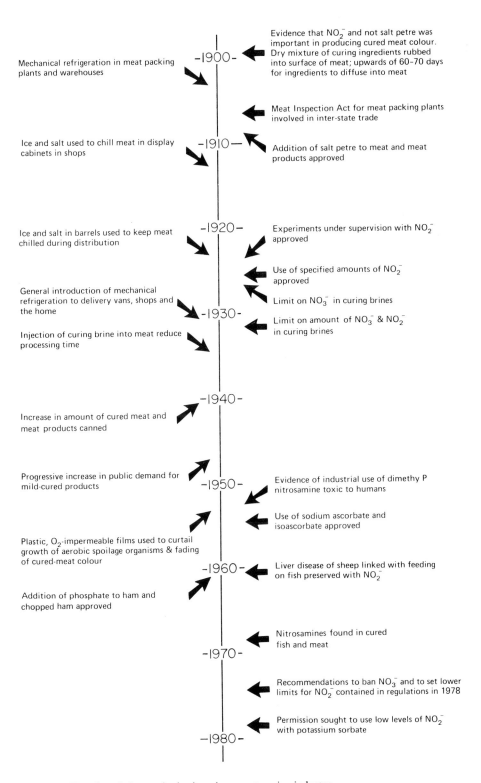

Fig. 9.8. Trends and changes in the American meat-curing industry.

(b) *Mild heat*

Although it is evident from Figure 9.8 that the meat industry has had several decades of experience in the manufacture of cured meat products that receive only moderate heat treatment before being introduced into the food distribution system, technological innovations can still pose novel problems. The heat-processing of luncheon meat (comminuted cured meat with non-meat binders such as starch) in relatively impermeable plastic (polyvinylidene chloride copolymer-PVDC) is an example. Tubes of PVDC are filled, the ends sealed with metal clips, cooked for 90 min at 70–75 °C (internal temperature in this range achieved after about 60 min), and chilled overnight. To ensure intimate contact of PVDC and the luncheon meat, the former was caused to shrink by immersion (10 s) in water at 93 °C. Bell and Gill (1982) noted that the freshly prepared product had an aerobic count of colony-forming units of about 10^2 g^{-1}, *Bacillus* and *Micrococcus* spp. occurring in approximately equal numbers. Storage at 10 °C for upwards of 42 d produced little if any change in the number of bacteria, glucose content or pH. At 25 °C, on the other hand, the *Bacillus* spp. proliferated at the surface but not within the luncheon meat, the failure at the latter site being attributed to the inhibitory effects of NO_2^- and NaCl being accentuated by a low redox. The surface growth was associated with acid drift in pH, accumulation of L(+)-lactic acid and glucose, the last-mentioned being produced by the breakdown of starch by amylase produced by the *Bacillus* spp. By the 14th day of storage, a *Streptococcus* sp. had emerged as the numerically dominant organism and there was a progressive accumulation of lactic acid. It was surmised that the *Bacillus* spp. pioneered this succession in two ways: they provided the streptococci with a fermentable substrate, glucose, and reduced the NO_2^- concentration to non-inhibitory levels.

Time–temperature combinations equivalent to 0.05–0.6 min at 250 °F ($F_0 = 0.05$–0.6 *cf.* $F_0 = 3$ for low acid foods, see p. 73) results in shelf-stable canned hams and meat products, providing refrigerated storage is used, even though vegetative non-thermoduric microorganisms only are killed. Indeed the good public health record of such products directed attention to factors that might be involved in the inhibition of either endospore germination or vegetative cell growth of *Clostridium botulinum* (see the next paragraph for details). A distinction must be drawn between the respective safety records of the products of commercial or home canning operations. Thus in the period 1899–1978, 51 outbreaks of botulism in Canada and the USA were attributed to meat and poultry products produced in the home but only 15 to those produced commercially. Of the latter, underprocessing was considered an important factor in 3 outbreaks and temperature abuse during storage in another 3. *Cl. botulinum* types A and B (minimum temperature of growth, 10 °C) were the only serotypes isolated. Much higher incidences of botulism associated with meat, principally pork, and poultry products have been recorded in certain

European countries. It has been assumed that this situation may be the result of war or economic depressions forcing the housewife to attempt home canning.

From a review of extensive literature, Christiansen (1980) put forward the following generalizations to account for the failure of *Clostridium botulinum* to grow in cured meat.

1 The nitrite level at the time of temperature abuse (*sic* when refrigerated storage is not used) is an important factor in determining the degree of inhibition.

2 Botulinal spores readily germinate in the presence of nitrite. However, nitrite inhibits by preventing outgrowth of germinated clostridial spores.

3 Nitrite levels decrease during storage; growth occurs when there is insufficient nitrite to check outgrowth.

4 Factors such as pH and ascorbate affect inhibition because they affect nitrite depletion.

5 Chelating agents—ascorbate, ethylenediaminetetra-acetic acid, cysteine—can influence the efficacy of nitrite by tying up iron.

6 Factors such as inoculum level, salt, temperature abuse and pH also affect the degree of inhibition because they directly affect growth of *Cl. botulinum*.

Since this summary was published it has been demonstrated that nitric oxide derived from NO_2^- probably inhibits the growth of nascent cells of *Cl. botulinum* (see (2) above) by reacting with the pyruvate—ferredoxin oxidoreductase in the phosphoroclastic reaction (Fig. 9.9).

Further to (5) above it has been shown, also, that *Clostridium botulinum* does not grow in cured meats containing low levels of NO_2^- (40 p/10^6) providing potassium sorbate (0.26%) is present.

These generalizations direct attention at the many interactions, some identified others hypothetical, that may influence nitrite inhibition of *Clostridium botulinum* and emphasize that little of value would be achieved by experiments in which A × B, A × C, A × D, ..., etc., interactions were studied in isolation (see also p. 6). Before considering the protocols of contemporary investigations, it is pertinent to consider the primary role of curing agents that have been shown to influence the action of NO_2^-.

(c) *Curing agents*

As the role of the various curing agents was discussed on p. 39, it will suffice to summarize the relevant information in the context of the growth of *Clostridium botulinum*.

(i) *Nitrite*. In the first decade of this century, Hoagland presented evidence that nitric oxide arising from nitrite—which in turn had been formed by reduction of saltpetre—reacted with myoglobin to give nitrosyl-myoglobin (see p. 39). Nitrite also acts as an antioxidant; it affects flavour, in part by preventing a 'warmed-up' flavour in cured meats, and may influence texture. It is a bacteriostatic agent, its toxicity increasing with

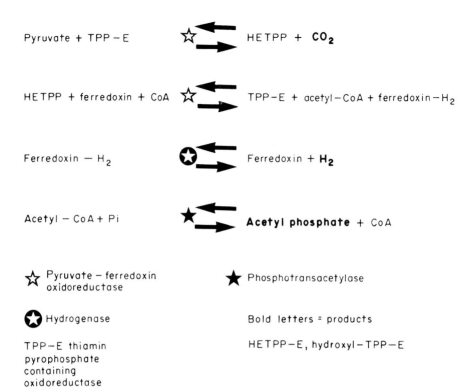

Pyruvate + TPP−E ☆ ⇌ HETPP + CO_2

HETPP + ferredoxin + CoA ☆ ⇌ TPP−E + acetyl−CoA + ferredoxin−H_2

Ferredoxin − H_2 ✪ ⇌ Ferredoxin + H_2

Acetyl − CoA + Pi ★ ⇌ **Acetyl phosphate** + CoA

☆ Pyruvate − ferredoxin oxidoreductase

✪ Hydrogenase

TPP−E thiamin pyrophosphate containing oxidoreductase

★ Phosphotransacetylase

Bold letters = products

HETPP−E, hydroxyl−TPP−E

Fig. 9.9. The growth inhibition of *Clostridium botulinum* in cured meats has been attributed to NO_2^- inhibition of the pyruvate-ferredoxin-oxidoreductase in the phosphoroclastic reaction.

increasing H-ion concentration. Nitrite reacts also with amines to produce nitrosamines (see p. 40).

(ii) *Ascorbate and isoascorbate.* These reductants and antioxidants react with NO_2^- to give nitric oxide thereby accelerating the rate of nitrosylmyoglobin formation; they also stabilize colour and flavour. When the ratio of ascorbate: nitrite is $> 2 : 1$, nitrosamine formation is inhibited. Through chelating iron, they may contribute also to the antimicrobial stability of cured meats.

(iii) *Polyphosphates.* Through enhancing water retention, these compounds are considered to aid tenderness, juiciness and flavour of cured meats. They also influence texture and, by chelating metal ions, act as antioxidants. The latter property might also contribute to the antimicrobial properties of cured meats. However, as polyphosphates have neutral reactions, they will tend to alleviate the inimical properties of NO_2^- by causing the pH of cured meats to be poised at or near a neutral reaction.

This summary draws attention to those secondary roles of curing agents and provides additional circumstantial evidence that the antibotulinal action of NO_2^- may well be influenced by substances that were chosen initially because of the beneficial organoleptic properties they contributed to cured meats.

The industrial use of dimethylnitrosamine led to research into the toxicity and carcinogenicity of nitrosamines. Their possible occurrence in foods was inferred from the study of death among sheep fed fish meal preserved with NO_2^-. Dimethylnitrosamine was shown to be present in the fish meal. In the following years, nitrosamines have been reported to occur in a variety of food commodities. When reviewing the literature on nitrosamines in various meat products, the conclusion was made (Anon 1980) 'that nitrosamines in very low concentrations may be found sporadically in various types of cured meat products'. This publication summarized the results of a detailed analysis of many cured products and identified one category requiring further research. In practice, therefore, the safety of cured meat involves the evaluation and balancing of two risks: illness due to botulism or disease due to nitrosamines. The trends in Fig. 9.8 show a gradual reduction in the permitted level of NO_2^- in the curing process, a reaction to the accumulation of scattered observations about the possible production of nitrosamines. Until curing can be achieved without recourse to NO_2^-, the critical concentrations of this substance for curing a meat as well as endowing it with antibotulinal properties are of cardinal importance. It was evident in the above discussions that the many interactions which occur between curing agents would defy analysis by simple experiments in which A was tested with B, with C, with D, etc.

Roberts and his colleagues (1981a) recognized the inadequacies of this simple approach and planned a study 'to acquire sufficient data to determine the minimum sodium nitrite concentration required to control *Clostridium botulinum* taking into account realistic commercial conditions of pH, NaCl, storage temperature, heat process and other additives such as sodium isoascorbate, polyphosphate and sodium nitrate.'

The general conclusions that could be based on their analysis of factors controlling the growth of *Clostridium botulinum* types A and B in pork slurries prepared from 'low' pH meat (pH range 5.5–6.3) are given below (see Roberts *et al.* 1981a, b for further discussion).

1 Increasing nitrite significantly reduced spoilage and toxin production. The relative effect of nitrite was smaller in combination with other significant factors such as 4.5% salt, isoascorbate, or storage at 15 °C but a combination of 200 μg g^{-1} nitrite with either 4.5% salt, or isoascorbate (1000 μg g^{-1}) or 15 °C storage always resulted in least spoilage and least toxin production.

2 Nitrate alone did not significantly affect spoilage but decreased toxin production. Chemical analysis indicated this to be a result of its reduction to nitrite. Several significant interactions involved nitrate.

3 Increasing salt concentration significantly decreased spoilage and toxin production. At high salt levels other significant factors, e.g. unit increase in nitrite concentration, had less effect (see (1) above).

4 Polyphosphate (Curaphos 700) did not affect spoilage overall but significantly increased toxin production in these 'low' pH slurries. This

increase in toxin production was counteracted by the addition of isoascorbate, or nitrate, or raising nitrite or salt concentration, or the heat treatment to the next highest level tested.

5   Isoascorbate significantly decreased spoilage and toxin production. In its presence the effect of many other significant factors (e.g. increasing salt, nitrite, heat treatment, storage temperature) were reduced. Spoilage and toxin production were least when isoascorbate was present in combination with 200 μg g^{-1} nitrite or 4·5% salt, or after high heat treatment or after storage at 15 °C.

6   High heat treatment significantly reduced spoilage and toxin production but little difference was observed between the effects of low and unheated levels.

7   The effect of storage temperature was significant and the lowest storage temperature (15 °C) resulted in least spoilage and toxin production. At low storage temperatures other significant factors (e.g. increasing salt or nitrite) were relatively less effective, although least spoilage and toxin production occurred when storage was at 15 °C in combination with 4.5% salt or 200 μg g^{-1} nitrite.

Taking this information as well as that obtained from studies of pork slurries obtained from 'high' pH meat (range 6.3–6.8), etc., Robinson *et al.* (1983) have derived a statistical model which allows an estimate of the probability of toxin production in the pork slurry system within *defined* limits.

The probability of toxin production is given by

$$P = \frac{1}{1 + e^{-\mu}}$$

where μ = the linear predictor.

For 'high' pH slurries, after low or high heat treatment, the improved model is:

$\mu = -7.647$

$-(1.603 \times N)$	where N = NaNO$_2$ (μg g^{-1} × 10^2)
$+(1.336 \times S)$	where S = NaCl (% w/v on the water)
$+(0.8748 \times T)$	where T = storage temperature (°C)
$-(2.134)$	if 500 μg g^{-1} NaNO$_3$ added
$-(6.468)$	if 1000 μg g^{-1} isoascorbate added
$-(1.789)$	if 0.3% w/v polyphosphate added
$-(1.839)$	if heat treatment HIGH
$-(0.01692 \times N \times T)$	
$-(0.01439 \times T^2)$	
$-(0.3599 \times S^2)$	
$+(0.6981 \times N)$	if 500 μg g^{-1} NaNO$_3$ added
$+(0.4289 \times N)$	if 0.3% w/v polyphosphate added
$-(1.295)$	if NaNO$_3$ *and* polyphosphate added
$+(0.8782 \times S)$	if isoascorbate added
$+(0.4192 \times N)$	if heat treatment HIGH
$+(1.02)$	if heat treatment HIGH *and* NaNO$_3$ added.

Models of this type (see Roberts *et al.* 1981c for a detailed discussion) would be expected to gain currency in the future thereby permitting detailed specification and monitoring of the screen (Fig. 9.10) that must be placed between the preparative and storage/distribution stages of food production and the consumer.

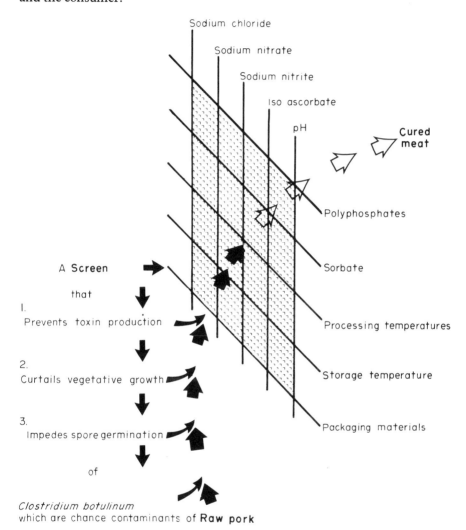

Fig. 9.10. The good safety record of cooked cured meats is due to the interaction of many factors.

(5) Codes of practice

From the above discussion, it was evident that one major 'screen' (e.g. Fig. 9.10) or, more commonly, several complementary 'screens' (Fig. 9.5) need to be set up between a consumer and one or several stages in food production and processing. Such 'screens' ought to ensure that bulk-produced foods or ingredients contain neither infective doses of the infectious type of food-poisoning bacteria nor toxins produced by organisms such as *Staphylococcus aureus*. If the 'screens' do not encompass methods that ensure a high probability of microbial death, then they ought to ensure very low levels of contaminating bacteria and methods of preservation/storage that hold contaminants in a quiescent state (viz. with *Bacillus cereus*, see p. 198). In rare instances only are such 'screens' likely to function until the time that a food has been finally prepared for a meal. Indeed, this stage is a common contributor to the events leading up to an outbreak of food poisoning and the one requiring general codes of practice rather than the more specific instructions that are required in a factory. Moreover, the former will often be implemented without rigorous supervision by trained persons. Before considering the material included in codes, it is pertinent to consider: (a) the principal findings of surveys that sought to identify the major factors that have contributed to outbreaks of food poisoning, and (b) a case history which illustrates the contribution of some of these factors.

(a) MAJOR FACTORS

From reports on 1152 outbreaks of food-borne disease that occurred in the USA between 1961–1976, Bryan (1978) noted that the following (given in descending rank order) were the main contributory causes*:
- (i)   inadequate cooling;
- (ii)   a lapse of a day or more between preparing and serving a meal;
- (iii) infected persons;
- (iv)   inadequate thermal processing, canning or cooling;
- (v)   inadequate hot storage;
- (vi)   inadequate reheating;
- (vii) ingesting contaminated *raw* food or ingredient;
- (viii) cross-contamination;
- (ix)   inadequate cleaning of equipment;
- (x)   obtaining food from unsafe sources;
- (xi)   using left-overs.

* See Roberts (1982) for details relating to UK.

(b) A CASE HISTORY

A charter flight (Passengers, 344; Crew, 20) landed at airport B where a new crew took over and the aircraft's galleys were provisioned with snacks and breakfasts (including cheese omelettes topped with slices of fried ham). The meals were prepared by a catering firm in the vicinity of airport B. The aircraft took off for airport C (estimated flight time, 8.5 h); the breakfasts were served approximately 7 h after departure (i.e. about 90 min before landing at airport C). A few passengers were taken ill shortly before landing and a large number within 1–2 h of arrival at airport C. In total 196 passengers (143 admitted to hospital) and 1 crew member became ill. The diagnosis based on clinical evidence (general malaise, vomiting, diarrhoea, abdominal pains, rapid onset of symptoms, of short duration in most cases) indicated that staphylococcal food poisoning was involved. This received support from laboratory examinations of uneaten breakfasts and the stools and vomitus of several patients; an enterotoxin-producing *Staphylococcus aureus* of phage group III was isolated. Enterotoxin was detected also in a sample of combined ham and omelette. An enterotoxin-producing *Staph. aureus* phage group III was isolated from the septic lesions on the fingers of one cook, the wrist of another and the nose of an assistant, all of whom were involved in the preparation of the omelettes topped with ham. Microscopic examination of remnants of ham revealed Gram-positive cocci in large numbers.

If the 11 contributory factors discussed above are used to analyse a schematic presentation (Fig. 9.11) of the main events in the preparation of the breakfasts that caused this outbreak of food poisoning, then 5 are seen to have been involved (bold roman numeral indicates the rank order of the factor in Bryan's (1978) analysis).

Inadequate cooling (**i**)—apart from the ham being held at room temperature throughout production (11.00–17.00 hours, day 2), poor temperature control of the finished product from 17.00 hours on day 2 to 07.00 hours on day 3 was followed by 7 h in galleys at the temperature of the aircraft's cabins.

Time lapse during preparation and serving of meal (**ii**)—in practice, upwards of 3 days had elapsed from the time that assistant No. 1 began preparing the ham for the breakfasts and the consumption of these by the passengers on the aircraft.

Infected persons (**iii**)—the circumstantial evidence indicates that the sores on the hands of cook No. 1 was the nidus of infection.

Inadequate reheating (**vi**)—reheating the breakfasts at 154 °C for 15 min obviously failed to destroy the enterotoxin-producing strain of *Staphylococcus aureus* phage group III as is evident from the recovery of the organism from the vomitus and stools of patients.

Cross-contamination (**viii**)—it would seem reasonable to presume that the plastic bucket used to store the fried ham was the vehicle that led to the organism from cook No. 1 being transferred to cook No. 2, assistant No. 2 and the ham.

The histogram in Fig. 9.11 shows not only that the outbreak among the 364 passengers and crew was explosive but, when considered as an epidemic curve, it has the profile characteristic of a common (or point) source outbreak in which all cases were exposed almost simultaneously to a common source of infection or intoxication, in this case the ham-topped omelettes. If, for the sake of argument, cook No. 2 had been involved with the preparation of a food that had been eaten by many people over a relatively long time interval, then a histogram having a different profile from that in Fig. 9.11 would have been obtained. In this latter event (a common source, multiple-event epidemic curve), cases would be expected to continue to occur at a high rate for a period equalling or exceeding the incubation period of staphylococcal food poisoning (Fig. 9.12).

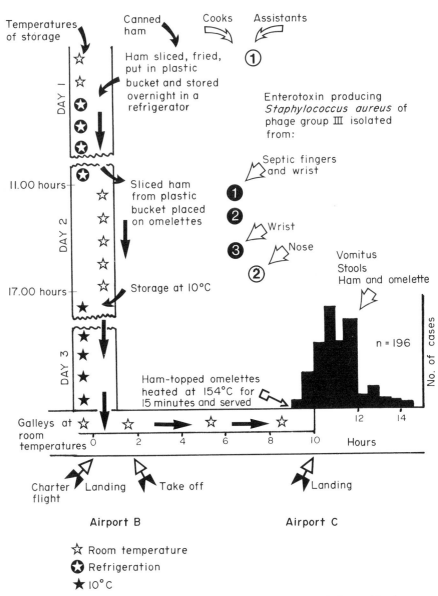

Fig. 9.11. A summary of the events contributing to an outbreak of staphylococcal food poisoning among the passengers on a charter flight.

CODES

It was evident in the above discussion of several causative agents of bacterial food poisoning that little of value would be achieved by one code that attempted to cover every contingency. It is preferable to consider codes that have been devised for particular foods and the specific methods of preparation. Moreover, the stringency or otherwise of a code will be dictated by a detailed analysis of the likely risk from particular types of food to the

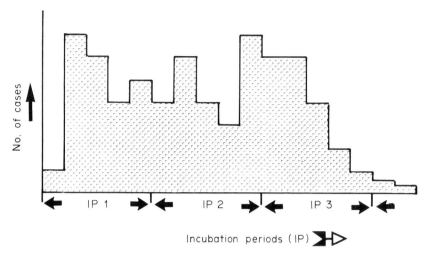

Fig. 9.12. When cases of food poisoning are due to multiple infection of a processed food, then the duration of the outbreak will be 2 or more times longer than the incubation period of the causative organism.

consumer (see Tables 9.5, 9.6 and 9.7). In other words, stringent codes are required for foods having a bad public health record and less stringent ones for those with little, if any, history of food poisoning. Codes for specific foods or classes of foods must emphasize important areas of control so that those who implement or supervise their implementation concentrate on pertinent features. In practice, of course, the preparation of codes and the education of those who are expected to observe them must be a joint exercise. When planning a code, the observations of Bryan (1978) will provide useful guidelines and effective control must be exercised with every factor listed below.

Temperature

The actual temperature-history of products under routine conditions of production, storage, distribution and preparation in the home or canteen must be given preference over temperatures that are assumed to obtain from periodic and unrecorded checks on the sensors fitted to cooling or heating systems.

Time

If for manufacturing or preparative purposes, foods have to be stored within the growth temperature range of microorganisms (Fig. 9.7), then the length of storage must match the dictates of an actual process rather than being determined by the whims of the cook.

Raw materials and ingredients

These must be checked thoroughly and all substandard or damaged materials discarded.

Preparation areas

A routine must be adopted so that raw and cooked foods do not make contact, either directly or indirectly via equipment.

Personal hygiene

The potential of a person to be a carrier, either temporal or chronic, of food-poisoning organisms such as *Salmonella* spp. or *Staphylococcus aureus* must be recognized and training must emphasize the requirement for minimal contact between an individual and an ingredient, a process or a food.

PIG BEL

If the customs of an ethnic group involve feasts in which a food notable as a vehicle for food poisoning, especially when prepared under unsatisfactory conditions, has a central role, then codes of practice may have little relevance. Indeed, active immunization of those at risk may well be the only preventive measure. Pig bel is an example of such a situation.

This disease, peculiar to Papua New Guinea, is caused by the β-toxin of *Clostridium perfringens* type C, an organism that is ubiquitously distributed in the soil and faeces of man and pigs. The toxin damages the mucosa, reduces the mobility of the villi and enhances bacterial attachment to the villi. Absorbed toxin causes necrosis of the intestinal wall underlying the affected mucosa and can lead to the death of the patient—commonly young children lacking antibodies to the toxin. Outbreaks are sporadic because they are associated with the consumption of large quantities of pork at festivals. As whole pigs are cooked in earth ovens, there is ample opportunity for contamination of the meat with *Cl. perfringens* and for its growth (Fig. 9.7) during the cooking/cooling phases. The normal diet also predisposes the people to infection. The low-protein staple diet depresses the secretion of proteolytic enzymes by the pancreas, and sweet potato, a common food, contains a trypsin-inhibitor which reduces the proteolytic activity of the pancreatic secretions. Indeed, the inhibition of proteases of pancreatic origin are considered to be prerequisites for outbreaks of pig bel because it protects the β-toxin from digestion.

(6) Recovery of food-poisoning organisms

The success of a routine monitoring programme for food-poisoning bacteria or investigations of outbreaks depends upon laboratory methods that permit the recovery of such organisms—or their products—from foods or environments in which non-pathogens are likely to be abundant. In addition, epidemiological studies depend upon refined methods of characterization, 'fingerprinting', in order to establish, for example, whether or not one or several depots or routes of infection are responsible for introducing a particular organism into a food chain or whether or not widely scattered outbreaks of food poisoning associated with a particular commodity have a common link to a depot or route of infection.

The summary of methods used to isolate food-poisoning bacteria (Table 9.11) directs attention to the use of a pre-enrichment stage—this provides damaged organisms with an opportunity to repair physiological lesions (see Table 3.1)—and an enrichment stage—this allows the out-numbered food-poisoning bacteria to achieve numbers similar to those of closely related non-pathogenic bacteria. Selective differential media are used so that the characteristic colony form of the microorganism sought can be recognized on a medium that commonly contains colonies of general contaminants. It will be appreciated that the schemes outlined in Table 9.11 are both time consuming and qualitative in nature. Indeed they provide an answer to one question only, Was a food-poisoning microorganism present in the sample of food examined? Failure to isolate such an organism is not

Table 9.11. An outline of methods used to isolate commonly occurring food-poisoning bacteria.

Bacteria	Method of isolation
Salmonella spp.	Pre-enrichment, e.g. in peptone water Enrichment in tetrathionate or selenite broth at 37 or 43 °C Isolation on Deoxycholate, Brilliant Green agar, etc.
Escherichia coli	Enrichment in Brain Heart Infusion broth for 3 h followed by incubation of a subculture in tryptone phosphate broth for 20 h at 44 °C Isolation on MacConkey agar
Yersinia enterocolitica	Enrichment in two stages: blended food in phosphate buffer incubated at 4 °C and subcultures made after 8, 14 and 21 d to a modified selenite medium with incubation at 23 °C Isolation on MacConkey agar
Vibrio parahaemolyticus	Enrichment in glucose Teepol broth for 18 h at 35 °C Isolation on thiosulphate citrate bile salts agar with incubation at 35 °C
Staphylococcus aureus	Enrichment in medium containing 10% NaCl not favoured currently because of failure of damaged cells to grow Isolation on Baird-Parker's medium
Clostridium perfringens	Isolation by direct-plating on a medium containing sulphite and cycloserine or neomycin and blood
Clostridium botulinum	Enrichment of heated (e.g. 80 °C for 30 to 60 min) and unheated samples in cooked meat medium at 25–30 °C Isolation on Blood agar or a medium containing egg yolk (lipase causes precipitation of the yolk)

proof of its absence. It may not have been included in the sample examined—a problem when contamination is contagious rather than random—there may have been minor faults with the media or maybe the task was done by an inexperienced operator. The methods outlined in Table 9.11 can be made semi-quantitative by using the Most-Probable-Number technique (see p. 146).

In epidemiological studies, the phenotypic properties selected by taxonomists for general classifications rarely permit the identification of strains within a species. Thus with *Salmonella*, biochemical characterization distinguishes three species, *Salmonella cholerae-suis*, *S. typhi* and *S. enteritidis*. If, however, the surface of the cell wall (Fig. 9.13) is examined serologically, then many major groupings (serogroups) of *Salmonella* can be

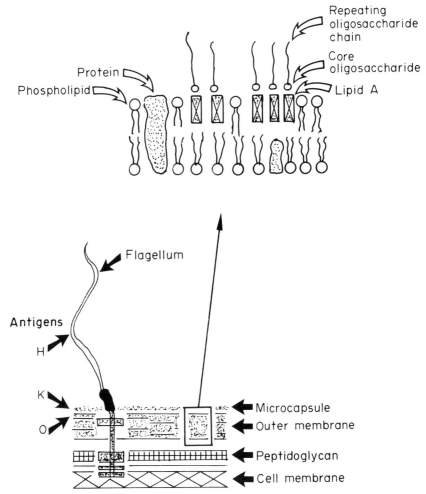

Fig. 9.13. The serotyping of enterobacteria is based on the antigens present in the flagellum or on the surface of the bacterial cell. The repeating units in the oligosaccharide chain (O antigens) on the surface of the cell and the H (flagella) antigens are used to type *Salmonella*. The microcapsule (K) antigens are used to type strains of *Escherichia coli*.

recognized. Further subdivision (serotypes) of the major groups can be achieved by the serological examination of the flagella. Thus with the Kauffmann and White scheme, an isolate can be given an antigenic formula viz:

$$\underbrace{\overset{\uparrow}{\text{Somatic or O antigens}}}_{1, 4, 5, 12} : \overset{\uparrow}{\underset{\downarrow}{1}} : \underbrace{1, 2}$$

Somatic or O antigens   Flagella (H) antigen in Phase 1

Flagella (H) antigen in Phase 2

In one system of nomenclature, emphasis is given to biochemical properties and serological analysis of the O and H antigens leads to the identification of serotypes viz: *S. enteriditis* serotype Typhimurium (antigenic formula as above). In another system, a serotype if accorded species rank, viz: *S. typhimurium*.

It will be appreciated that serotyping, as applied to *Salmonella*, provides a valuable system of 'fingerprinting' for epidemiological purposes. When it is recalled that *Salmonella typhimurium* was, until recently, the most commonly isolated organism of this genus, it will be realized that serotyping was of limited value in tracing transmission routes or identifying depots of infection. Phage typing permits further subdivision of this serotype and resistance to antibiotics (an indirect way of analysing an organism's content of plasmids) can allow even further discrimination between strains of a serotype. Figure 9.14 illustrates the use of various typing procedures in a study of the movement of *S. typhimurium* between calf rearing units and the human population in the UK. Table 9.12 lists methods that are used to characterize food-poisoning bacteria for epidemiological studies.

Table 9.12. Typing methods used in epidemiological studies of food-poisoning bacteria.

	Serotyping		Toxin neutralization test	Phage typing	Antibiotic sensitivity test
	Cell	Toxin			
Salmonella	+	−	−	+	+
Clostridium botulinum	−	+	+ (1 ng)*	−	−
Staphylococcus aureus	−	+	−	+	−
Clostridium perfringens	−	+	−	−	−
Vibrio parahaemolyticus	+	−	−	−	−
Bacillus cereus	+	−	−	−	−

* Lower limit of detection. +, Used routinely; −, not used routinely.

(7) Sampling plans

This topic was discussed on p. 55.

216

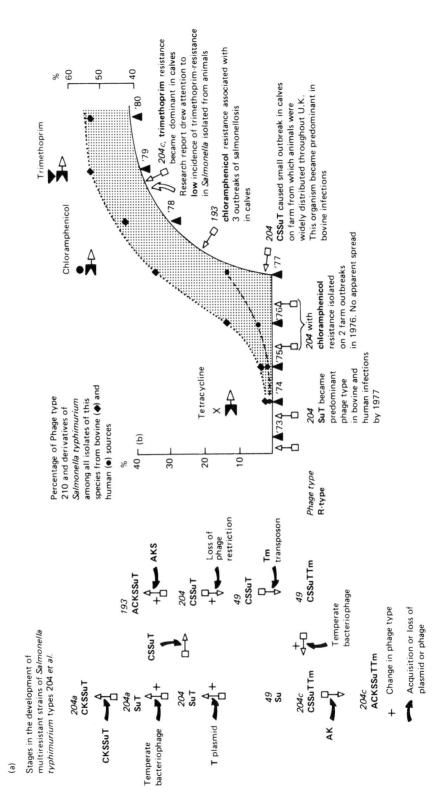

(a)

Stages in the development of multiresistant strains of *Salmonella typhimurium* types 204 et al.

(b)

Percentage of Phage type 210 and derivatives of *Salmonella typhimurium* among all isolates of this species from bovine (◆) and human (●) sources

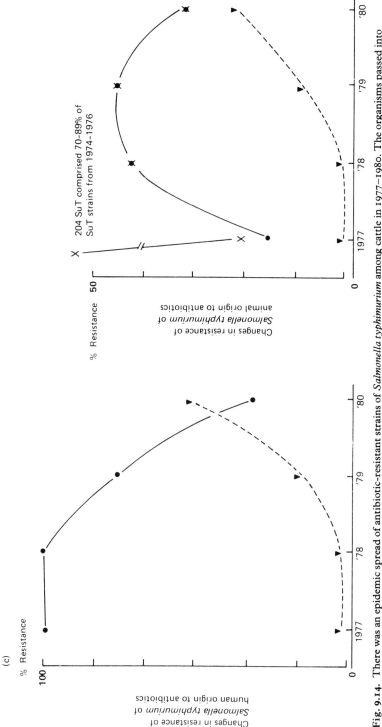

Fig. 9.14. There was an epidemic spread of antibiotic-resistant strains of *Salmonella typhimurium* among cattle in 1977–1980. The organisms passed into the food chain and caused infections in humans. This Figure summarizes (a) the acquisition of antibiotic resistance by the organisms, their spread (b) through the cattle population, and their entry (c) into the food chain. For references to these events see Trelfall (1983). Drug resistance symbols: A, ampicillin; C, chloramphenicol; K, neomycin–kanamycin; S, streptomycin; Su, sulphonamides; T, tetracylines; Tm, trimethoprim.

217

FUNGAL POISONING AND MYCOTOXINS

The general definition of food poisoning at the beginning of this chapter encompasses illness resulting from the ingestion of plant materials containing a constitutive or induced toxic substance or animal material in which such substances had been acquired from a food chain (examples are given in Table 9.1). It would seem reasonable to assume that early man learnt by trial and error to identify toxic plants and animals of his habitat and that such a learning process would need to be repeated following migration. Indeed, it is of interest to note the misfortunes of the sailors of Van der Hagen, an explorer who visited a new habitat, the Mauritius, in 1607. They caught with ease the Pink Pigeons (*Nesoenar mayeri*) that abounded on the island but suffered 'a prostration and dejection in all limbs' when they ate the birds. The silors were unaware that the tameness of the pigeons—and their own illness—was caused by a toxin, a narcotic present in the seeds of the Fangame tree (*Euphorbia pyrijolia*) upon which the birds fed. Even today a thorough knowledge of fungal identification is essential if poisoning due to the ingestion of certain mushrooms is to be avoided—the blackening of silver by poisonous mushrooms is a most dubious test. People still die through incorrect identification of mushrooms; in Europe, for example, upwards of 90% of deaths are due to the ingestion of *Amanita phalloides*, the death cap fungus. Poisoning is not indicated until several hours following ingestion, abdominal pain is followed by diarrhoea and severe emesis. If dehydration of a patient is treated clinically, the symptoms subside and recovery appears imminent but death often occurs on the third day because of damage to the liver and kidney. Several toxic substances, mainly cyclic polypeptides, have been isolated from *A. phalloides*. A phallotoxin, phalloidin, impairs liver function through damage to the cell membrane in experimental animals. An amotoxin, alpha-amanitin, which damages nuclei is probably the principal toxin. With this example—and several others could be cited—the ingestion of a recognizable entity, a mushroom, is followed by well-characterized clinical symptoms and pathological changes in specific organs.

In contrast to fungal food poisoning, a muddled picture has emerged during the past twenty years with respect to mycotoxins and human mycotoxicoses. Chemical and pharmacological investigations of fungal products in food, feeds and laboratory media have provided a large body of information about toxic metabolites. As yet, however, the paucity of epidemiological evidence about the aetiological role of the majority of these metabolites precludes an assessment of their contribution to human illness. Indeed, Busby and Wogan (1979), for example, were forced to adopt an artificial, hierarchical classification system when attempting to marshal the available evidence. In practice they recognized four categories of toxin (Table 9.13), viz:

1   those that have been identified chemically in food;
2   those that are produced by fungal isolates from foods;

Table 9.13. *Mycotoxins*

Aflatoxins

AFG₁

AFG₂

AFB₁

AFB₂

Name and toxins derived from *Aspergillus flavus*. Four major toxins; B_1 & B_2 fluoresce blue and G_1 & G_2 green with UV light. A hydroxylated product of AFB is secreted in the milk of cattle ingesting feed containing AFB. Liver damage is a major symptom in animals fed toxins. Carcinogenic properties also demonstrated

Zearalenone

Produced by *Fusarium* spp. in which it probably acts as a hormone in regulating formation of perithecia. Isolated from corn and a range of cereals. Has oestrogenic properties; pigs fed contaminated feed exhibit hyperoestrogenic symptoms

	R_1	R_2	R_3	R_4
R_1		H_2	=O	H
H_2				

(continued)

Table 9.13.—*Continued*

Ochratoxin A

R	R_1	R_2
H	Cl	H

Produced by *Aspergillus ochraceous* commonly found in certain cereals—wheat, oats and rye—and corn. Pigs fed infected feed develop nephropathy

Patulin

Produced by *Penicillium expansum*. Detected in apples, cider and apple juice. Subcutaneous sarcomas induced in rats

Sterigmatocystin

R	R_1	R_2	R_3
H	CH_3	H	H

Produced by *Aspergillus versicolor*, for example, and structurally related to aflotoxins. Detected in wheat and green coffee beans. Induce liver cancer in rats

Trichothecenes

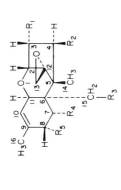

Acyl- or hydroxyl-substituted

Produced by a range of fungi, e.g. *Fusarium* and *Stachybotrys*. Found in cereal grain and hay. A cytotoxin that can cause necrosis of the skin or damage to the epithelial mucosa of the stomach and small intestine

221

3 those that have been identified in animal feeds or in cultures of organisms isolated therefrom;

4 those that have not as yet been implicated with foods or feeds.

The following mycotoxicoses belong to category (1).

(a) Ergotism, either gangrenous (muscular pain, prickly sensation in limbs which become swollen, inflamed, necrotic and eventually slough off) or convulsive (cramps, severe convulsions leading to death), is produced by ergot alkaloids (amide derivatives of D-lysergic acid), smooth muscle stimulants produced by *Claviceps purpurea* growing on the grains of cereals. Wet springs followed by hot summers with intermittent rainy periods favours the organism's life cycle. The germination of sclerotia which have over-wintered on the soil is promoted by a wet spring, the dispersal of ascospores from the fruiting body formed by the sclerotia is favoured by dry windy conditions and the germination of ascospores on the pistils of flowering grain is aided by warm wet weather. Hyphae that invade the kernel of the grain develop sclerotia which tend to be most toxic when young. This may well account for the epidemic outbreaks of ergotism that used to follow famines, a population being forced to use freshly harvested grain containing young sclerotia.

(b) Alimentary toxic aleukia is a mycotoxicosis that reached epidemic proportions in the USSR during and immediately following the 1939–45 war. Mould-infected grain of wheat and prosomillet that had overwintered in the field was the major cause; the grain being eaten by people who could not obtain properly stored cereal. During overwintering, *Fusarium* and *Cladosporium* became important contaminants of cereal grains and laboratory studies established that toxin production by *Fusarium poae* and *Fusarium sporotrichoides* was maximal at low incubation temperatures or in cultures exposed to freezing and thawing cycles. Of several toxic materials isolated from these moulds, trichothecenes (Table 9.13) appear to be of major importance in the disease syndrome.

(c) Endemic (Balkan) nephropathy, a fatal disease in which atrophy of renal tubules is a characteristic feature, is thought to be caused by ochratoxin A. This toxin tends to be more prevalent in endemic than the non-endemic areas of the Balkans. Evidence suggests, also, that rain at particular stages in the growing cycle of maize in endemic areas predisposes stored grain to attack by *Aspergillus ochraceous*.

(d) Aflatoxicosis. Epidemiological studies of the estimated incidence of liver cancer and intake of aflatoxins is beginning to provide a circumstantial link between the two. The toxin is produced by *Aspergillus flavus*.

The following diseases also are probably mycotoxicoses of category 1 or 3 (causative organisms and toxins given in parenthesis): red mould disease (*Fusarium* spp.; trichothecenes?) caused by ingestion of grain from infected crops; Stachybotryoxicosis (*Stachybotrys atra*; trichothecenes), a disease of farm workers exposed to contaminated hay; cardiac beriberi (*Penicillium citreoviride*; citroviridin) associated with infection of rice grains.

References and Further Reading

The following Journals are important primary sources of information:

Journal of Applied Bacteriology
Journal of Food Technology
Journal of Dairy Science
Journal of Dairy Research
Food Technology (the *Overview* series in this publication is strongly recommended.)
Journal of Food Science

Agricultural and Biological Chemistry
Journal of the Science of Food and Agriculture
Journal of Hygiene, Cambridge
The Journal of Food Protection
European Journal of Applied Microbiology and Biotechnology
Systematic and Applied Microbiology

The following review series are useful sources of information:

Davies R. (Ed.) (1982) *Developments in Food Microbiology 1*. Applied Science Publishers, Barking.

Advances in Applied Microbiology
Microbiological Reviews
Society for Applied Bacteriology—Technical series
Society for Applied Bacteriology—Symposium series
Progress in Industrial Microbiology
Advances in Food Research

Chapter 1

Brown M.H. (Ed.) (1982) *Meat Microbiology*. Applied Science Publishers, Barking.

Gould G.W. & Corry J.E.L. (Eds) (1980) *Microbial Growth and Survival in Extremes of Environment*. Academic Press, London.

Mossel D.A.A. & Ingram I. (1955) The physiology of the microbial spoilage of foods. *J. Appl. Bact.* **18**, 232–68.

Russell A.D. & Fuller R. (Eds) (1979) *Cold-tolerant Microbes in Spoilage and the Environment*. Academic Press, London.

Skinner F.A. & Roberts T.A. (Eds) (1983) *A Symposium on Food Microbiology*. Academic Press, London.

The International Committee on Microbiological Specifications for Foods (1980) *Microbial Ecology of Foods 1. Factors affecting life and death of microorganisms*. Academic Press, London.

Chapter 2

Crosby N.T. & Sawyer R. (1976) *N*-nitrosamines: a review of chemical and biological properties and their estimation in foodstuff. *Adv. Food Res.* **22**, 1–72.

Davies R., Birch G.G. & Parker K.J. (Eds) (1976) *Intermediate Moisture Foods*. Applied Science Publishers, Barking.

Dryden F.D. & Birdsall J.J. (1980) Why nitrite does not impart colour. *Food Tech.* **34** (July) 29–42.

Duckworth R.B. (Ed.) (1975) *Water Relations of Foods*. Academic Press, London.

Friend J. & Rhodes M.J.C. (Eds) (1982) *Recent Advances in the Biochemistry of Fruit and Vegetables*. Academic Press, London.

Kochan I. (1973) The role of iron in bacterial infections with special consideration of host–tubercle *Bacillus* interaction. *Curr. Topics Microbiol Immunol.* 60, 1–30.

Koryeka-Dahl M.B. & Richardson T. (1978) Activated oxygen species and oxidation of food constituents. *CRC Critical Reviews in Food Science and Nutrition* 10, 209–42.

Kramer A., Solomos, T., Wheaton F., Puri A., Sirivichaya S., Lotem Y., Fowke M. & Ehrman L. (1980) A gas exchange system for extending the shelf life of raw foods. *Food Tech.* 34 (July) 65–74.

Law B.A. (1979) Reviews of the progress of dairy science: Enzymes of psychrotrophic bacteria and their effects on milk and milk products. *J. Dairy Res.* 46, 573–88.

Livingstone D.J. & Brown W.D. (1981) The chemistry of myoglobin and its reactions. *Food Tech.* 35 (May) 244–52.

Mead G.C. & Freeman B.M. (Eds) (1980) *Meat Quality in Poultry and Game Birds*. Longmans, London.

Report (1972) *Food Additives and Contaminants Report on the Review of the Perservatives in Food Regulations, 1962*. HMSO, London.

Roberts R.A., Hobbs G., Christian J.H.B. & Skovgaard O.O. (Eds) (1981) *Psychrotrophic Microorganisms in Spoilage and Pathogenicity*. Academic Press, London.

Rockland L.B. & Stewart G.F. (1981) *Water Activity: influences on food quality*. Academic Press, London.

Skinner F.A. & Hugo W.B. (Eds) (1976) *Inhibition and Inactivation of Vegetative Microbes*. Academic Press, London.

Tilbury R.H. (Ed.) (1980) *Developments in Food Preservatives 1*. Applied Science Publishers, Barking.

Willis R.H.H., Lee T.H., Graham D. & McGlasson W.B. (1981) *Post-harvest: An Introduction to the Physiology and Handling of Fruit and Vegetables*. Granada, London.

Overview: Extending the shelf life of fresh foods by combining controlled atmosphere and refrigeration (1980) *Food Tech.* 34 (March), 44–71.

Overview: Changing technologies/changing microbiological problems (1980) *Food Tech.* 34 (October), 58–84.

Overview: Food ingredient safety review programs (1981) *Food Tech.* 35 (December), 68–87.

Chapter 3

McLaughlin T. (1969) *Cleaning Hygiene and Maintenance Handbook*. Business Books Ltd., London.

Sharpe A.N. (1980) *Food Microbiology: A Framework for the Future*. Charles C. Thomas, Springfield, Illinois.

The International Committee on Microbiological Specifications for Foods (1978) *Microorganisms in Foods 1. Their significance and methods of enumeration*, 2e. Academic Press, London.

The International Committee on Microbiological Specifications for Foods (1974) *Microorganisms in Foods 2. Sampling for Microbiological Analysis: Principles and specific applications*. University of Toronto Press, Toronto.

Overview: Microbiological standards for food—an update (1978) *Food Tech.* 32 (January), 51–67.

Overview: Rapid methods and automation for food microbiological analysis. (1979) *Food Tech.* 33 (March), 51–74.

Overview: Quality control—a function of quality assurance strategy (1981) *Food Tech.* 35 (April), 47–67.

Overview: Analysis of constituents in foods for improvement of processes (1983) *Food Tech.* 37 (January), 74–79.

Chapter 4

Beuchat L.R. & Rice S.L. (1979) *Byssochlamys* spp. and their importance in processed foods. *Adv. Food Res.* 25, 237–88.

Hersom A.C. & Hulland E.D. (1980) *Canned Foods: Thermal Processing and Microbiology*. Churchill Livingstone, London.

Hoyem T. & Kvale O. (Eds) (1977) *Physical, Chemical and Biological Changes in Food Caused by Thermal Processing*. Applied Science Publishers, Barking.

Jackson J.M. & Shinn B.M. (1979) *Fundamentals of Food Canning*. Avi Publishing Co., Westport, Connecticut.

Josephson E.S. & Peterson M.S. (Eds) (1983) *Preservation of Food by Ionizing Radiation*. CRC Press Inc., Florida.

Lampi R.A. (1977) Flexible packaging for thermoprocessed foods. *Adv. Food Res.* **23**, 306–428.

Put H.M.C., van Doren H., Warner W.R. & Kruiswijk J.T. (1972) The mechanism of microbiological leaker spoilage of canned foods: a review. *J. Appl. Bact.* **35**, 7–27.

Put H.M.C., Vitwoet H.J. & Warner W.R. (1980) Mechanisms of microbiological leaker spoilage of canned foods: biophysical aspects. *J. Food Prot.* **43**, 488–97.

Stumbo C.R. (1973) *Thermobiology in Food Processing*. Academic Press, London.

Stumbo C.R., Purohit K.S., Ramakrishnan T.B., Evans D.A. & Francis F.J. (1983) *CRC Handbook of Lethality Guides for Low-Acid Canned Foods*. CRC Press Inc., Florida.

Urbain W.M. (1978) Food irradiation. *Adv. Food Res.* 24, 155–228.

Food radiation features (1983) *Food Tech.* 37 (February), 38–60.

Overview: Application of quantitative analysis to sterilization processes (1978) *Food Tech.* **32** (March), 59–83.

Overview: Thermal processing of canned foods (1978) *Food Tech.* 32 (June), 53–76.

Overview: Food quality improvement through kinetic studies and modelling (1980) *Food Tech.* **34** (February), 51–95.

Chapter 5

Carr J.G., Cutting C.V. & Whiting G.C. (1975) *Lactic Acid Bacteria in Beverages and Food*. Academic Press, London.

Hurst A. (1981) Nisin. *Adv. Appl. Microbiol.* 27, 85–123.

Kinsella J.E. & Hwang D.H. (1976) Enzymes in *Penicillium roqueforti* involved in the biosynthesis of cheese flavour. *Crit. Rev. Food Sci. Nutrition.* 8, 191–228.

Pederson C.S. (1979) *Microbiology of Food Fermentations*, 2e. Avi Publishing Co., Westport, Connecticut.

Rhodes-Roberts M.E. & Skinner F.A. (Eds) (1982) *Bacteria and Plants*. Academic Press, London.

Robinson R.K. (Ed.) (1981) *Dairy Microbiology, vol. 1. The microbiology of milk*. Applied Science Publishers, Barking.

Robinson R.K. (Ed.) (1981) *Dairy Microbiology, vol. 2. The microbiology of milk products*. Applied Science Publishers, Barking.

Rose A.H. (Ed.) (1982) *Fermented Foods*. Academic Press, London.

Scott R. (1981) *Cheese Making Practice*. Applied Science Publishers, Barking.

Skinner F.A., Passmore S.M. & Davenport R.A. (Eds) (1981) *Biology and Activities of Yeasts*. Academic Press, London.

Skinner F.A. & Quesnel L.B. (Eds) (1978) *Streptococci*. Academic Press, London.

Overview: Microbial alterations in high-protein foods (1978) *Food Tech.* **32** (May), 173–98.

Overview: Use of microbial cultures to increase the safety, shelf life and nutritive value of food products (1981) *Food Tech.* **35** (January), 70–94.

Chapter 6

Garvie E.I. (1978) *Int. J. Syst. Bact.* 28, 190–193.

Mossel D.A.A. & Ingram I. (1955) The physiology of the microbial spoilage of foods. *J. Appl. Bact.* **18**, 232–68.

Skinner F.A. & Lovelock D.W. (Eds) (1979) *Identification Methods for Microbiologists*. Academic Press, London.

The International Committee on Microbiological Specifications for Foods (1980) *Microbial Ecology of Foods 2. Food Commodities*. Academic Press, London.

Overview: Update on food spoilage organisms (1979) *Food Tech.* 33 (January), 55–87.

Chapter 7

Anon (1978) Waterborne infectious disease in Britain. *J. Hyg., Camb.* 81, 139–49.

Cundell A.M. (1981) Rapid counting methods for coliform bacteria. *Adv. Appl. Microb.* **27**, 185–205.

Dean R.B. (1981) *Water Re-use : Problems and Solutions.* Academic Press, London.

Lewis M.J. (1983) Editorial: the bacteriological examination of drinking water. *J. Hyg., Camb.* **90**, 143–147.

Twort A.G., Hoather R.C. & Law F.M. (1974) *Water Supply.* Edward Arnold, London.

Chapter 8

Benefield L.D. & Randall C.W. (1980) *Biological Process Design for Waste Water Treatment.* Prentice-Hall, Englewood Cliffs, N.J.

Curds C.R. (1982) The ecology and role of protozoa in aerobic sewage treatment processes. *Ann. Rev. Microbiol.* **36**, 27–46.

Dean R.B. & Lund E. (1979) *Water Re-use : Problems and Solutions.* Academic Press, London.

Hobson P.N., Bousfield S. & Summers R. (1981) *Methane Production from Agricultural and Domestic Waste.* Applied Science Publishers, London.

Sundstrom D.W. & Klei H.E. (1979) *Waste Water Treatment.* Prentice-Hall, Englewood Cliffs, N.J.

Vesiland P.A. (1979) *Treatment and Disposal of Waste Water Sludges.* Ann Arbor Science, Michigan.

Winkler M. (1981) *Biological Treatment of Waste Water.* Ellis Horwood, Chichester.

Yaziz M.I. & Lloyd B.J. (1982) The removal of *Salmonella enteriditis* in activated sludge. *J. Appl. Bact.* **53**, 169–72.

Chapter 9

Anon (1980) A survey of nitrosamines in sausages and dry cured meat products. *Food Tech.* **34** (July) 45–53.

Anon (1982) Epidemiology: Food poisoning and Salmonellosis surveillance in England and Wales (1981). *Br. med. J.* **285**, 1127–8.

Ball A.P. *et al.* (1979) Human botulism caused by *Clostridium botulinum* type E: the Birmingham outbreak. *Q. J. Med., New Series* **48**, 473–91.

Bell R.G. & Gill C.O. (1982) Microbial spoilage of luncheon meat prepared in an impermeable plastic coating. *J. Appl. Bact.* **53**, 97–102.

Bryan F.L. (1978) Factors that contribute to outbreaks of food-borne disease. *J. Food Prot.* **41**, 816–27.

Bryan F.L. (1981) Current trends in food-borne Salmonellosis in the United States and Canada. *J. Food Prot.* **44**, 394–402.

Bryan F.L. (1981) Hazard analysis of food service operations. *Food Tech.* **35** (February), 78–88.

Busby W.F. & Wogan G.R. (1979) Chapter 11. In *Food-borne Infections and Intoxications*, 2e (Eds Reimann H. & Bryan F.L.). Academic Press, London.

Christiansen L.N. (1980) Overview. *Food Tech.* **34** (May), 237–9.

Corry J.E.L., Roberts D. & Skinner F.A. (Eds) (1982) *Isolation and Identification Methods for Food Poisoning Organisms.* Academic Press, London.

Coulston F. (Ed.) (1979) *Regulatory Aspects of Carcinogenesis and Food Additives : the Delaney Clause.* Academic Press, London.

Gibson A.M., Roberts T.A. & Robinson A. (1982) *J. Food Technol.* **17**, 471.

Hobbs, B.C. & Christian J.H.B. (Eds) (1973) *The Microbiological Safety of Foods.* Academic Press, London.

Hobbs G. (1976) *Clostridium botulinum* and its importance. *Adv. Food Res.* **22**, 135–86.

Imbey C.S., Mead G.C. & George S.M. (1982) Competitive exclusion of *Salmonella* from chicken caeca using a defined mixture of bacterial isolates from the caecal microflora of an adult bird. *J. Hyg., Camb.* **89**, 479–490.

Kornacki J.L. & Marth E.H. (1982) Food-borne illness caused by *Escherichia coli* : A review. *J. Food Prot.* **45**, 1051–67.

Lewis G.E. (Ed.) (1982) *Biomedical Aspects of Botulism.* Academic Press, London.

226

Lorenz K. & Hoseney R.C. (1979) Ergot on cereal grains. *CRC Critical Reviews in Food Science and Nutrition* **11**, 311–54.

Pohl P. & Leunen J. (Eds) (1982) *Resistance and Pathogenic Plasmids.* National Institute for Veterinary Research, Brussels.

Recheigh M. (1982) *CRC Handbook of Food Borne Diseases of Biological Origin.* CRC Press Inc., Florida.

Reiman H. & Bryan F.L. (Eds) (1979) *Food-borne Infections and Intoxications,* 2e. Academic Press, London.

Roberts D. (1982) Factors contributing to outbreaks of food poisoning in England and Wales 1970–1979. *J. Hyg., Camb.* **89**, 491–498.

Roberts T.A., Gibson A.M. & Robinson A. (1981a) Factors controlling the growth of *Clostridium botulinum* A and B in pasteurized, cured meats. I. *J. Food Technol.* **16**, 267.

Roberts T.A., Gibson A.M. & Robinson A. (1981b) Factors controlling the growth of *Clostridium botulinum* A and B in pasteurized, cured meats. II. *J. Food Technol.* **16**, 239.

Roberts T.A., Gibson A.M. & Robinson A. (1981c) Prediction of toxin production by *Clostridium botulinum* in pasteurized pork slurry. *J. Food Technol.* **16**, 337.

Roberts T.A., Gibson A.M. & Robinson A. (1982) *J. Food Technol.* **17**, 307.

Robinson A., Gibson A.M. & Roberts T.A. (1982) *J. Food Technol.* **17**, 727.

Skirrow M.B. (1982) Special article: *Campylobacter* enteritis—the first five years. *J. Hyg., Camb.* **89**, 175–84.

Sofos J.W., Busta F.F. & Allen C.E. (1979) Botulism control by nitrite and sorbate in cured meats: a review. *J. Food Prot.* **42**, 739–70.

Swaminathan B., Harmon M.C. & Mehlman I.J. (1982) A review: *Yersinia enterocolitica. J. Appl. Bact.* **52**, 151–83.

Tartakow I.J. & Vorperian J.H. (1981) *Food-borne and Water-borne Diseases: their Epidemiologic Characteristics.* Avi Publishing Co., Westport, Connecticut.

Trelfall E.J., Frost J.A., King H.C. & Rowe B. (1983) Plasmid encoded trimethoprim resistance in salmonellas isolated in Britain between 1970 and 1980. *J. Hyg., Camb.* **90**, 55–60.

Overview: *Clostridium perfringens* food-borne illness (1980) *Food Tech.* **34** (April), 76–95.

Overview: An assessment of nitrite for the prevention of botulism (1980) *Food Tech.* **34** (May), 228–57.

Overview: Food-borne pathogenic bacteria of emerging significance (1982) *Food Tech.* **36** (March), 71–96.

Overview: Current problems in botulism of interest to the food industry (1982) *Food Tech.* **36** (December), 87–118.

Index